# WOODY PLANTS OF KENTUCKY AND TENNESSEE

# Woody Plants of Kentucky and Tennessee

### THE COMPLETE WINTER GUIDE TO THEIR IDENTIFICATION AND USE

Ronald L. Jones and B. Eugene Wofford

UNIVERSITY PRESS OF KENTUCKY

Copyright © 2013 by The University Press of Kentucky

Scholarly publisher for the Commonwealth, serving Bellarmine University,
Berea College, Centre College of Kentucky, Eastern Kentucky University,
The Filson Historical Society, Georgetown College, Kentucky Historical Society,
Kentucky State University, Morehead State University, Murray State University,
Northern Kentucky University, Transylvania University, University of Kentucky,
University of Louisville, and Western Kentucky University.
All rights reserved.

*Editorial and Sales Offices:* The University Press of Kentucky
663 South Limestone Street, Lexington, Kentucky 40508-4008
www.kentuckypress.com

17 16 15 14 13     5 4 3 2 1

Library of Congress Cataloging-in-Publication Data
Jones, Ronald L. (Ronald Lee)
  Woody plants of Kentucky and Tennessee : the complete winter guide to their identification and use / Ronald L. Jones and B. Eugene Wofford.
       p. cm.
  Includes bibliographical references and index.
  ISBN 978-0-8131-4250-0 (hardcover : alk. paper) -- ISBN 978-0-8131-4309-5 (epub)
ISBN 978-0-8131-4310-1 (pdf)   1. Woody plants--Kentucky--
Identification.   2. Woody plants--Tennessee--Identification.   3. Woody
plants--Kentucky--Pictorial works.   4. Woody plants--Tennessee--Pictorial
works.   5. Plants in winter--Kentucky--Identification.   6. Plants in
winter--Tennessee--Identification.   7. Plants in winter--Kentucky--
Pictorial works.   8. Plants in winter--Tennessee--Pictorial works.   I.
Wofford, B. Eugene. II. Title.
  QK162.J67 2013
  582.1609768--dc23                                                      2013018805

This book is printed on acid-free paper meeting
the requirements of the American National Standard
for Permanence in Paper for Printed Library Materials.

Manufactured in the United States of America.

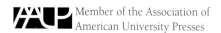 Member of the Association of
American University Presses

To my brother, Thomas Eugene (Tommy) Jones, in fond memory of our many happy days in the fields and woodlands of Middle Tennessee.   *RLJ*

My contribution to this project began shortly after receiving an organ transplant in March 2005. With sincere and humble gratitude I dedicate this book to the following who helped in so many ways during this difficult time and made my survival possible: to my wife, Deborah; my colleague, David K. Smith; my physicians, Steve Silver and Santiago Vera; and especially to an anonymous organ donor!   *BEW*

# The Financial Support of the Following Contributors Is Gratefully Acknowledged

Austin Peay State University

Breedlove, Dennis & Associates, Inc.

Copperhead Environmental Consulting

Eastern Kentucky University Department of Biological Sciences and Office of Graduate Education and Research

Eco-Tech Consultants

Kentucky Native Plant Society

Kentucky Society of Natural History

Southern Appalachian Botanical Society

Tennessee Native Plant Society

Third Rock Consultants

University of Tennessee–Department of Ecology & Evolutionary Biology

University of Tennessee–Office of Research

# Acknowledgments

The following individuals provided fresh specimens for photographic use: Edward W. Chester, Aaron Floden, Dennis Horn, W. Michael Dennis, Dwayne Estes, D. K. Smith, Ross Clark, Zack Murrell, Derick Poindexter, Paul Durr, Ken McFarland, Pat Cox, Jamey McDonald, Joe Feeman, David Webb, Meredith Clebsch, Ed Clebsch, John Beck, William Martin, Ed Schilling, Pine Ridge Gardens, Growild, Inc., and Theo Witsell.

Funding for field support was provided by the Hesler Fund and the Breedlove-Dennis Fund, The University of Tennessee, Knoxville. Zack Murrell, Appalachian State University, E. Wayne Chester, Austin Peay State University, and J. Richard Abbott, Missouri Botanical Garden, reviewed drafts of the manuscript and their comments and corrections are much appreciated. Many undergraduate and graduate students at Eastern Kentucky University were involved in the field testing and proofing of the keys, and their assistance was a great help in detecting typographical errors and other problems with the keys. We would like to thank our wives, Kathleen Jones and Deborah Wofford, for all their patience and understanding throughout this project, and especially for their long-term support, both personally and professionally.

# A Note on the Photographs and Figures

All color images included in this book were taken by B. Eugene Wofford. Images were made with a Nikon D70s digital camera and an AF micro nikkor 60-mm lens attached to a Polaroid MP4+ photo stand equipped with two side-mounted 300-W halogen lamps positioned at ca. 45 degrees from the photo stand platform. In order to eliminate shadows, specimens were placed on a sheet of Plexiglas removed from a "2x2 slide box viewer" and backed with a sheet of herbarium paper. Images were placed in Adobe Photoshop 7.0 only to enhance contrast and brightness; no color changes were made. With very few exceptions, all images were made from freshly collected specimens between October and March 2005–2012. An additional 200+ images are available at the following website: http://tenn.bio.utk.edu/vascular. This online set contains additional images of some species depicted in the text, as well as images of some species not included in the text (mostly from groups such as *Crataegus* and *Amelanchier* in which there is little difference in winter features among the species).

Chris Fleming, BDY Environmental LLC, constructed figure 1 and the generalized geographic coverage map on the back cover. Graham Sexton provided technical assistance to figure design.

The cover photograph of ice-covered mountain laurel was taken by Marty Silver on Unaka Mountain, January 22, 2011.

# Contents

- LIST OF FIGURES .................................................................................................................... x
- **SECTION I. INTRODUCTION**
  - PURPOSE ........................................................................................................................ 1
  - COVERAGE ...................................................................................................................... 1
  - HOW TO USE THIS BOOK ............................................................................................... 2
  - SOURCES OF INFORMATION .......................................................................................... 2
  - FORMAT AND ABBREVIATIONS ...................................................................................... 3
  - MORPHOLOGICAL FEATURES FOR WOODY PLANTS IN WINTER ................................... 4
  - GYMNOSPERMS VS. ANGIOSPERMS .............................................................................. 15
  - USES OF WOODY PLANTS FOR FOOD, MEDICINE, FIBER, AND WEAPONS ................... 18
  - CONSERVATION CONCERNS ......................................................................................... 20
  - DISCLAIMER .................................................................................................................. 20
- **SECTION II. KEY TO KEYS** ................................................................................................. 21
- **SECTION III. KEYS TO GENERA** ........................................................................................ 23
- **SECTION IV. GENERIC AND SPECIES ACCOUNTS** ............................................................ 45
- **SECTION V. PLATES 1 TO 630** ........................................................................................... 117
- LITERATURE CITED .............................................................................................................. 119
- APPENDIX I. GLOSSARY ....................................................................................................... 121
- APPENDIX II. WOODY PLANTS USEFUL AS FOOD ............................................................. 125
- APPENDIX III. WOODY PLANTS USEFUL FOR MEDICAL NEEDS ....................................... 127
- APPENDIX IV. WOODY PLANTS USEFUL FOR CORDAGE .................................................. 129
- APPENDIX V. WOODY PLANTS USEFUL FOR CONSTRUCTING BOWS AND ARROWS ..... 131
- INDEX OF SECTION I ............................................................................................................ 133
- INDEX OF SCIENTIFIC NAMES IN SECTIONS IV AND V ..................................................... 135
- INDEX OF COMMON NAMES IN SECTION IV .................................................................... 141

# FIGURES

| | | |
|---|---|---|
| FIGURE 1. | PHYSIOGRAPHIC REGIONS OF KENTUCKY AND TENNESSEE | 4 |
| FIGURE 2. | MAJOR TWIG FEATURES (LABELED) | 5 |
| FIGURE 3. | LEAF ARRANGEMENTS, PITH CONDITIONS, AND ARMATURE | 6 |
| FIGURE 4. | ARMATURE, CORKY RIDGES, AND TWIG SCARS | 7 |
| FIGURE 5. | BUD TYPES, AERIAL ROOTS, AND TENDRILS | 8 |
| FIGURE 6. | FRUITS TYPES (ACHENE TO FOLLICLE) | 9 |
| FIGURE 7. | FRUITS (HIP TO SAMARA) AND SEED CONES | 10 |
| FIGURE 8. | INFLORESCENCE TYPES | 11 |

# SECTION I. INTRODUCTION

## PURPOSE

This book provides a guide to the identification and uses of woody plants of the Kentucky–Tennessee (KY–TN) region during winter conditions. Included are diagnostic keys to the genera and species, color photographs that emphasize the characteristic features, descriptions of the genera, and species accounts that include scientific name, common name, habitat, distribution, conservation status for rare species, invasiveness categories for non-native species, and wetland affinities. Additional notes are provided on plants that can potentially be used for food, medicine, fiber, and weapons during the winter, including lists of the best plants for these purposes.

The information in this book should be of interest to a cross-section of citizens of Kentucky, Tennessee, and surrounding regions, including professional botanists and consultants, high school teachers and students, foresters and farmers, agricultural agents, land planners and landscape architects, and many others. Those desiring to learn more about the practical uses of wild plants in winter, especially during emergency survival situations, will also find much of interest in this book.

## COVERAGE

This book is limited to woody plants—trees, shrubs, and woody vines. Woody plants are defined as those plants with stems that persist above the ground through the winter because of the development of secondary tissues (bark and wood). Another defining characteristic found in nearly all woody plants is the presence of over-wintering above-ground buds (with *Polygonella* being a notable exception). Three groups of woody plants can be defined, somewhat arbitrarily, as follows: **trees** have one or a few self-supporting main stems and heights of over 7 meters; **shrubs** have self-supporting multiple stems and heights of less than 7 meters; and **woody vines** have woody stems that cannot support their own weight; that is, they climb on other vegetation and man-made structures or sprawl or spread on the ground. Herbaceous plants (those in which the above-ground parts die back each year) are therefore excluded, but some slightly woody plants (as in *Rubus*), as well as subshrubs (low-growing, herblike woody perennials) such as species of *Epigaea*, *Gaultheria*, and *Mitchella*, are included.

Only native and naturalized non-native species are included. Native species are defined as those that were present (to the best of our knowledge) in the region at the time of European settlement. Naturalized non-native species are therefore those species that are native outside of the KY–TN region, either to other parts of North America or to other continents, having been introduced either deliberately or accidentally, that are able to reproduce in the region and have now become established as a part of the flora. Many non-native species, such as Douglas-fir (*Pseudotsuga menziesii*) and English white oak (*Quercus robur*), are cultivated in the region but are not able to reproduce and become naturalized. Only those non-native species that we consider to be truly naturalized (our general rule of thumb was the documentation of the species in more than five counties in the region) are included in the text.

This book provides an account of 172 genera (142 native and 30 non-native) and 457 species and lesser taxa (381 native and 76 non-native) in Kentucky and Tennessee. Of these 457 taxa, 15 are gymnosperms and 442 are angiosperms. Of the angiosperms, 10 are monocots and 432 are dicots.

The geographic boundaries for this book are the state boundaries of Kentucky and Tennessee. In reality, this book has a much broader application, as most of the species covered have much wider ranges than Kentucky and Tennessee (see map on back cover). The vast majority of the woody species of West Virginia, western Virginia–North Carolina–South Carolina, northern Georgia–Alabama–Mississippi, eastern Missouri–Arkansas, and southern Indiana–Illinois–Ohio are included in this text. The physiographic regions (see figure 1) covered by this book include the Blue Ridge, Valley and Ridge, Appalachian Plateaus, Interior Low Plateaus, and northern Coastal Plain (Mississippi Embayment).

## How to Use This Book

To identify a species using this book, the following method is recommended. **First**, become familiar with the botanical terminology used in the keys. A glossary and color photographs with terms identified are provided in this text. Additional information on botanical terminology of woody plants is available in related books (Jones 2005, Lance 2004, Swanson 1994, Wofford and Chester 2002), and there are also many websites that provide extensive information on botanical terminology. **Second**, obtain a 10× hand lens or similar means of magnification (especially useful for examining bundle trace scars). **Third**, use the sequential sets of keys (Sections II, III, and IV) to identify the genus and species. Section II provides keys to the major groupings (Keys A–N). Section III provides keys to genera, and Section IV provides keys to species along with detailed species accounts. All the keys are dichotomous, with two lead statements—always read both leads of a particular couplet before making a decision as to which route to take (read completely both #1s, make a selection as to which best describes the plant in question, then read both #2s, and so on). In cases where the genus is known but the species is in question, go directly to the Generic and Species Accounts (Section IV), in which the genera are arranged alphabetically, and key to species. **Fourth**, check the photograph(s) provided for the species in Section V and determine if there is a match. If not, then backtracking may be necessary, or the complete keying process tried again.

Every attempt has been made to make the keys user-friendly, particularly by eliminating technical terms where possible and by providing more detail in the keys and making the keys more descriptive. In some cases the term AND or OR is used to emphasize contrasting choices in the keys, and these couplets, in particular, require careful reading to avoid being led astray. Figures 2 through 8 provide color images of botanical terms frequently used in the text. Appendix I is a glossary of botanical terms. Appendixes II through V provide listings of woody plants that can be used for food, medicine, cordage, and weapons. Color photographs of all genera and about 90% of the species in the text are provided in Section V. Three indexes are found at the back of the book. The first is an index to Section I; the second is an index of scientific names; and the third is an index listing common names.

## Sources of Information

A number of other texts and resources were consulted in the writing of this book. Distributions of taxa are based on county maps provided in Clark and Weckman (2008) and Chester et al. (1993, 1997), as well as the continually updated website of the University of Tennessee Herbarium (http://tenn.bio.utk.edu/vascular). Many references were consulted in the construction of keys, including Brown and Brown (1972), Cope (2001), Campbell et al. (1975), Core and Ammons (1958), Harlow (1946), Lance (2004), Preston and Wright (1978), and Swanson (1994). Primary references for information on the uses of woody plants for fiber, food, medicine, and weapons were: Department of the Army (2003), Burrows and Tyrl (2001), Elias and Dykeman (1982), Krochmal and Krochmal (1984), Lewis and Elvin-Lewis (1977), Meuninck (2007, 2008), Moerman (1998), Muenscher (1975), Peterson (1977), Thayer (2006), and Westbrooks and Preacher (1986). The nomenclature (scientific and common names) used in the text chiefly follows that used in Jones (2005) and Chester et al. (2009). In a few cases the scientific names have been updated to reflect recent nomenclatural changes in the literature. For some families two names are given—in these cases the first family name is the one that is the more widely used and recognized, and the second name is one that is more recently proposed and less widely used. Author citations are based on Brummitt and Powell (1992). Latin meanings, chiefly based on Gledhill (2008), are given for genera and species except where the meaning is self-evident, such as *alabamensis*, *americana*, *canadensis*, *caroliniana*, *mexicana*, *virginiana*, and so on. Also, the most commonly used color-related epithets are listed here and not repeated later: *alba* (white), *flava* (yellow), *nigra* (black), and *rubra* (red).

Listings maintained by the Exotic Plant Councils of Kentucky and Tennessee (TN-EPPC 2009; KY-EPPC 2011) are the bases for the inclusions of Severe, Significant, and Lesser Threat categories in the accounts of non-native species.

Wetland indicators, based on USDA, NRCS (2011), are defined as follows: OBL—obligate wetland plant, over 99% chance of occurring in wetlands; FACW—facultative wetland plants, 67 to 99% chance of occurring in wetlands; FAC—facultative plants, 34 to 66% chance of occurring in wetlands; FACU—facultative upland plants, 1 to 33% chance of occurring in wetlands; UPL—upland plants, less than 1% chance of occurring in wetlands, and NI—no indicator due to insufficient information. Plus (+) and minus (−) signs are used to indicate an increased or decreased tendency to occur in wetlands.

## Format and Abbreviations

Each generic account has the following information: scientific name, common name, family name, Latin meaning of generic name, habit (tree, shrub, or vine), bark features, twig features (including pith type, bud positions and types), leaf scar arrangement, bundle scar number, presence or absence of stipular scars, fruit type, and other notes. If leaves are evergreen, then arrangement and leaf features are given and detailed twig and bud features may be omitted. Unless otherwise noted in the generic description, the species can be assumed to be deciduous, the twigs with continuous pith, and leaf scars in more than one plane. Also included in the generic accounts is information on the uses of the species for foods, medicines, fiber, and weapons, and whether or not the species are known to be poisonous.

Each species account has the following information: scientific name, common name, habitat, KY and TN distribution, frequency, area of origin if non-native, rarity status in each state, invasiveness status, wetland status, Latin meaning of specific epithet, and other notes of interest. Two wetland indicator abbreviations are given when the status is different in KY (in Wetland Region 1) and TN (in Wetland Region 2) but only one when it is the same in both states. The final entry is a reference to the plate number(s) that provide images of the species.

Distributional abbreviations used in the species accounts are as follows: for regional sections of each state—W (western), C (central), E (eastern), N (northern), and S (southern); for physiographic provinces—AP (Appalachian Plateaus of KY and TN); BG (Bluegrass region of KY), BR (Blue Ridge of TN), CB (Central Basin of TN), CP (Coastal Plain of KY and TN), IP (Interior Low Plateaus of KY and TN), and VR (Valley and Ridge of TN). Standard two-letter abbreviations are used for states.

Several other abbreviations are used frequently in the text: an exclamation point (!) at the beginning of the species account is used to indicate a rare species listed at the state and/or federal levels (Crabtree 2012, KSNPC 2010); an asterisk (*) to indicate a non-native species; a double asterisk (**) to indicate a non-native species listed as a Severe, Significant, or Lesser Threat in KY or TN (or Alert in TN); a skull and bones (☠) to indicate poisonous parts; a thumb's up (👍) to indicate edible parts; a medical symbol (⚕) to indicate a plant with medicinal uses; and an arrow (➺) to indicate a plant with fiber or weapon uses.

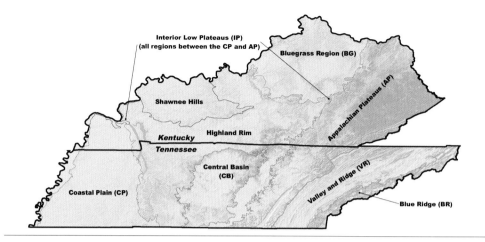

FIGURE 1 / PHYSIOGRAPHIC REGIONS OF KENTUCKY AND TENNESSEE.

# MORPHOLOGICAL FEATURES FOR WOODY PLANTS IN WINTER

Winter identification of woody plants can be a daunting and frustrating exercise, but we believe this guide and an understanding of critical morphological features will enable users to identify all woody plants in our region to the genus level and in most cases to the species level. During the growing season woody plants offer innumerable vegetative, floral, and fruiting characteristics to aid in identification, but winter identification is much more challenging, with leaves, flowers, and fruits usually absent. A few species, such as witch-hazel (*Hamamelis virginiana*) and mistletoe (*Phoradendron leucarpum*), flower in late fall or winter, but in the vast majority of species there is little or no evidence of flowers or their remnants. Fruits, while being absent in most species, do sometimes persist, especially in the case of species that produce dry-type fruits, such as capsules, follicles, legumes, and nuts, and their presence is often very helpful in determining the identity of the plant. Fleshy fruits deteriorate quickly with the onset of freezing temperatures, and are usually not evident in the winter. Species in about 50 genera in the KY–TN region produce evergreen leaves, and these are relatively easy to identify compared to the deciduous-leaved species. For most people it is the leaves that provide the most easily recognized features for plant identification, and their absence in deciduous-leaved species often poses a major stumbling block. In fact, the vast majority of deciduous-leaved species can be quite readily identified in winter by using easily observable features of the twig (buds, leaf scars, bundle scars, and stipular scars) and bark. A few other features may also be present for those willing to do a little extra detective work, such as checking for odor or taste or cutting the twig to examine the pith or searching for remnant fruits and leaves, which may be still attached to a few branches or lying about on the ground in the vicinity of the plant, and these kinds of clues often provide the clincher for an identification. The important morphological characteristics for winter identification are described next. See figures 2a and 2b for a general overview of the parts of a twig, and figures 3 through 8 for color illustrations of many of the morphological features mentioned. The **Glossary** (Appendix I) provides a comprehensive listing of terms used in the text.

INTRODUCTION 5

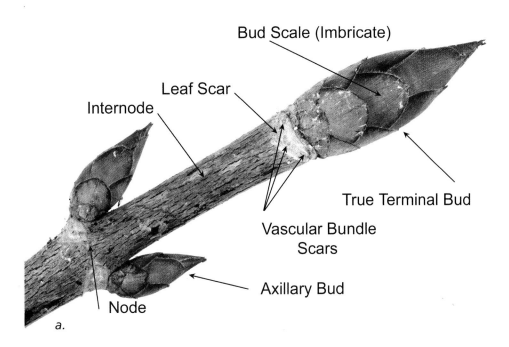

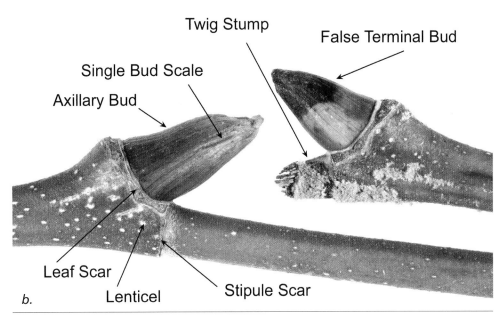

**FIGURE 2 / MAJOR TWIG FEATURES (LABELED).** A. TWIG WITH TRUE TERMINAL BUD (*AESCULUS GLABRA*), AND B. TWIG WITH FALSE TERMINAL BUD (*PLATANUS OCCIDENTALIS*).

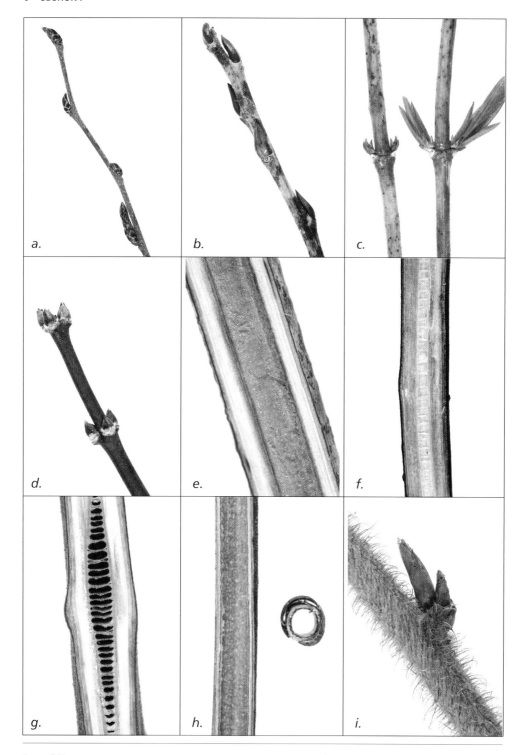

FIGURE 3 / LEAF ARRANGEMENT, PITH CONDITIONS, AND ARMATURE. A. ALTERNATE IN 1 PLANE, B. ALTERNATE IN > 1 PLANE, C. OPPOSITE, D. WHORLED, E. PITH CONTINUOUS, F. PITH DIAPHRAGMED, G. PITH CHAMBERED, H. PITH EXCAVATED, I. TWIG WITH BRISTLES.

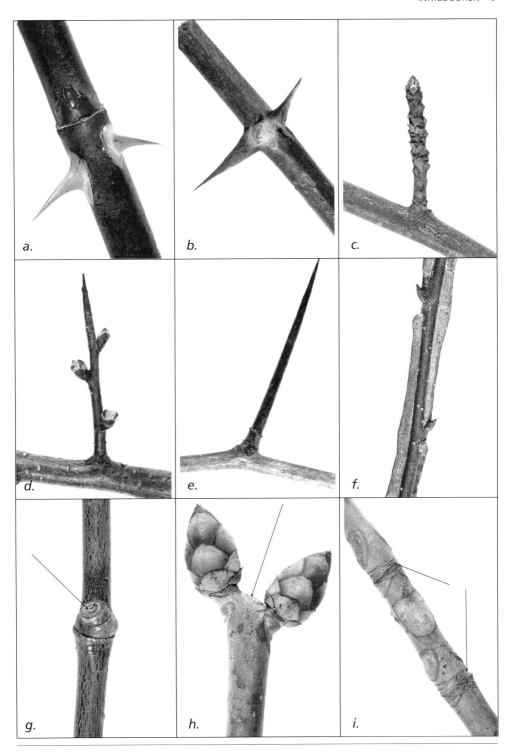

**FIGURE 4 / ARMATURE, CORKY RIDGES, AND TWIG SCARS.** A. PRICKLE, B. STIPULAR SPINE, C. SPUR SHOOT, D. THORN WITH BUDS, E. THORN LACKING BUDS, F. CORKY RIDGES, G. BRANCH SCAR, H. FRUIT SCAR, I. BUD SCALE SCARS.

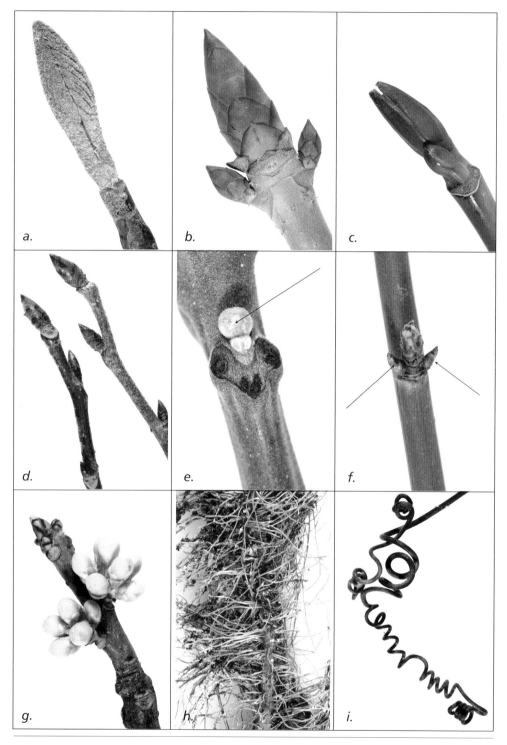

**FIGURE 5 / BUD TYPES, AERIAL ROOTS, AND TENDRILS.** A. NAKED (LACKING SCALES), B. TRUE TERMINAL BUD WITH IMBRICATE SCALES, C. TRUE TERMINAL BUD WITH VALVATE SCALES, D. FALSE TERMINAL BUD WITH IMBRICATE SCALES IN TWO ROWS, E. SUPERPOSED BUD, F. COLLATERAL BUDS, G. FLOWER BUDS, H. AERIAL ROOTS, I. TENDRILS.

**Figure 6 / Fruit types (achene to follicle).** A. Achene, B. Aggregate of follicles, C. Aril, D. Berry (inferior ovary), E. Berry (superior ovary), F. Capsule (inferior ovary), G. Capsule (superior ovary), H. Drupe (inferior ovary), I. Follicle.

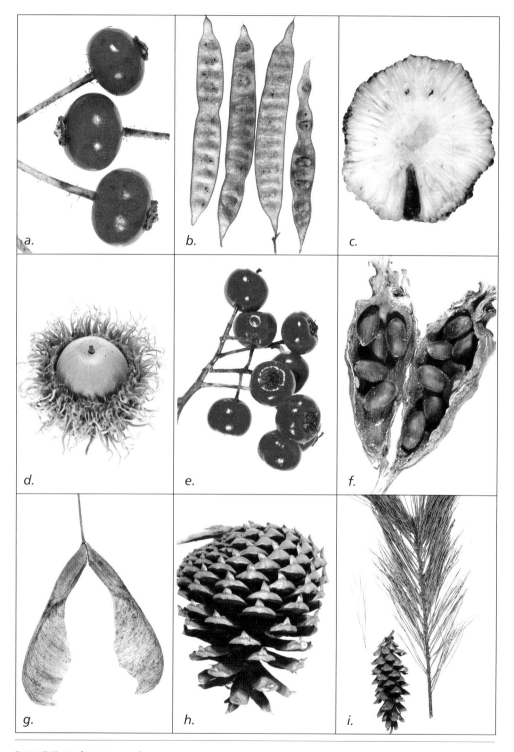

**FIGURE 7 / FRUITS (HIP TO SAMARA) AND SEED CONES.** A. HIP, B. LEGUME, C. MULTIPLE, D. NUT, E. POME, F. PSEUDOCARP, G. SAMARA, H. SEED CONE (HARD PINE), I. SEED CONE (SOFT PINE).

**Figure 8 / Inflorescence types.** A. Female catkin, B. Male catkin, C. Corymb of capsules (inferior ovaries), D. Cyme of capsules (superior ovaries), E. Head (or capitulum, with cypsela fruits), F. Panicle of drupes (superior ovaries), G. Raceme of capsules (superior ovaries), H. Spikelets (caryopsis fruits), I. Umbel of berries (inferior ovaries).

## Leaves

Leaves provide many important identifying features in evergreen species as well as in many deciduous-leaved species that retain their leaves for much of the winter (described here as remnant leaves). Truly evergreen leaves are typically bright green, glossy, and leathery, and leaves will be persistent on the older twigs, not just the twigs of the previous year's growth. Remnant leaves are more membranous and usually turn brownish or yellowish in winter. The major leaf features related to identification are described here.

Leaves may occur in alternate, opposite, or whorled arrangements. Stems with **alternate** leaves produce one leaf per node, and alternate leaves may be two-ranked, that is, in **one plane** (see figure 3a) with 180° between adjacent leaves, or in several ranks, that is, in **more than one plane** (see figure 3b). Stems with **opposite** leaves (see figure 3c) produce two leaves at each node, that is, on opposite sides of the stem from each other. In the **whorled** leaf arrangement (see figure 3d), there are three or more leaves per node, all attached in a whorl at the same general area of the twig. In the **Leaf Scars** section there is additional discussion on the distinguishing of alternate leaves in one plane from alternate leaves in more than one plane.

Leaf blades may be simple or compound. If simple, the leaf blade is not completely divided into distinct units, but it may be variously toothed or lobed. If compound, the leaf blade is divided into several distinct leaflets. The leaflets may be arranged pinnately (like the bristles of a feather) or palmately (like the fingers of the hand). Trifoliolate refers to a compound leaf with three leaflets. The flattened portion of a leaf is the **blade**, and the stalk of a leaf is a **petiole**. The stalk of a leaflet in a compound leaf is a **petiolule**, and the axis of a compound leaf is a **rachis**. Special terms (see **Glossary**) are associated with the different kinds of leaf tips (**apices**), edges (**margins**), and lower sides (**bases**).

For examples of these leaf descriptions, see the images in Section V of *Elaeagnus pungens* (for alternate, simple leaves), *Lonicera japonica* (for opposite, simple leaves), *Chimaphila maculata* (for whorled, simple leaves), *Mahonia bealei* (for leaves alternate and pinnately compound), *Sibbaldiopsis tridentata* (for leaves alternate and trifoliolate), and *Clematis terniflora* (for leaves opposite and pinnately compound).

## Twigs

The twig is the terminal portion of a branch produced during the most recent year's growth. Twigs provide the most conspicuous morphological features necessary for identification, and they may also have distinctive color, taste, and odor. Twigs may have conspicuous **lenticels**—small, somewhat corky features that may be found on an otherwise completely smooth twig surface. They may be dotlike, elongated slits, either vertical or horizontal, or variously irregular. Except for their presence and degree of prominence they are of little value in identification. The **node** of a twig is the location of a leaf attachment, and an **internode** is the portion of the twig between nodes. Lenticels, nodes, and internodes are illustrated in Figures 2a and 2b.

## Twig Scars

There are a variety of scars that can be found on twigs in the winter, and these can be defined as follows:

1. *Branch and Fruit Scars*. Twigs that die back in some species will result in the formation of false terminal buds (discussed later in this section). In addition, fallen branches (see figure 4g) and fruit stalks (see figure 4h) may also produce scars. These scars and their bundle scars tend to be round in shape. They will also differ from leaf scars in size, shape, and/or bundle scar architecture. Except for a few species, as in *Pyrularia pubera* with its diagnostic branch scars, these features are of little value for identification.

2. *Leaf Scar.* The scar remaining on the twig where the leaf was attached is the leaf scar (see figure 2a); a corky abscission layer forms at the petiole base prior to leaf fall and persists afterward. The shape of the leaf scar is sometimes helpful in identification. Twigs with leaf scars in **ONE PLANE** (see figure 3a) have the scars aligned in two rows (2-ranked or distichous), one on each side of the twig; twigs with leaf scars in **MORE THAN ONE PLANE** (see figure 3b) have the scars aligned in many ranks (or spiraling) along the twig. To distinguish these two differences, point the terminal end of the twig straight toward your eyes and observe the alignment of the buds and leaf scars.
3. *Stipule Scars.* Stipules are leaflike structures attached to the twig adjacent to the leaf petiole (see figure 2b); stipules vary in size and degree of attachment and upon falling they leave a broad range of scars, from a tiny dot (*Salix*) or slit (*Ilex*) to those that nearly (*Fagus*) or completely (*Magnolia*) surround the twig. In a few cases they are attached along the base of the petiole (*Rubus*) and leave no scar on the twig.
4. *Vascular Bundle Scars.* These are the scars (or scar) remaining within the leaf scar of the broken end(s) of vascular (xylem and phloem) conducting tissue passing from the stem to the leaf (see figure 2a); they are highly variable in number, arrangement, and shape. The number, shape, and arrangement of bundle scars are often helpful in identification. Most frequently these scars number one, three, or many (five or more).
5. *Bud Scale Scars.* These scars occur at the position where the end buds were located at the beginning of the previous season's growth (see figure 4i); they are of little assistance in identification, but do provide a means of aging the twig.

## Buds

These structures are dormant during winter and contain the growing point of the stem and next year's leaves and flowers. In a few cases, flower buds are formed in late fall and also remain dormant during winter. Buds produce numerous morphological features important for winter identification.

Buds are usually visible but in a few cases they are sunken into the twig and not readily distinguishable. Most species have buds that are surrounded by relatively hard protective **scales** (see figure 2a) that are typically modified leaves or, very rarely, stipules. Some species have buds unprotected by scales—these are called **naked buds** (figure 5a). They closely resemble modified folded leaves, usually with visible venation and/or various types of dense pubescence (trichomes or "hairs"). In species with scaly buds, the number of scales is commonly two or more, with one-scaled buds present only in a few genera, such as *Magnolia* and *Salix*. When a bud has two visible scales that meet along the edges but do not overlap (duckbill-like), the condition is **valvate** (figure 5c). If the bud scales are more than two and overlapping, the condition is termed **imbricate** (see figure 2a), and this is the most frequently encountered kind of bud. Imbricate buds may have the scales arranged in several rows or in two rows (see figure 5b and 5d).

Buds are of two types: **terminal** and **lateral**. A **true terminal bud** (see figure 2a) is at the exact tip of the twig. In some cases, however, a portion of the twig withers and dies along with the true terminal bud. The uppermost remaining lateral bud will then assume the position of the true terminal bud, thus producing a **false terminal bud** (see figure 2b). A false terminal bud is slightly offset at the apex of the twig and can be distinguished by the presence of both a leaf scar and a twig scar (or the stump of the fallen twig) immediately beneath or adjacent to the false terminal bud.

The classification of lateral buds depends upon their position in relation to the leaf scar. **Axillary** buds (see figure 2b) are formed in the leaf axis directly above the leaf scar. If more than one bud is found above the leaf scar, the one closest to the scar is the axillary bud; the other(s) is termed **superposed** (see figure 5e). **Collateral** buds (see figure 5f) are produced to the right and/or left of the axillary bud. Flower buds may be present in some species (see figure 5g). A **catkin** is a specialized flower bud present during winter (see figure 8a and 8b); it is a dense spike or raceme of unisexual flowers in a few genera (*Alnus, Betula, Corylus*, etc.). These often open during warm periods in early spring. With the exception of *Hamamelis* (witch-hazel), only a very few other woody species flower in winter and early spring, including species of *Acer, Neviusia, Syringa,* and *Ulmus*.

## Pith

The pith occupies the central portion of the twig and is composed of thin-walled non-conducting storage tissue. In cross-section it may be round or variously lobed or angled. It is commonly white but may be variously colored, mostly either brown, yellow, or pinkish. In longitudinal section, several morphological variations may exist: (1) **continuous**: pith is homogeneous and continuous (see figure 3e); (2) **diaphragmed**: continuous pith with horizontal plates at regular intervals (see figure 3f); (3) **chambered**: small empty spaces when the pith is absent between the diaphragms (see figure 3g); (4) **spongy**: pith containing cavities or pores like a sponge; and (5) **excavated**: when pith is absent or essentially so (see figure 3h)—that is, the twig is hollow.

## Armature, Corky Ridges, and Climbing Adaptations

Woody plants may produce a variety of sharp-pointed, stiff, or bristly features: (1) **bristle**: flexible, non-spiny epidermal outgrowths (see figure 3i); (2) **prickle**: sharp-pointed epidermal outgrowths easily broken off (see figure 4a); (3) **stipular spine**: sharp-pointed stipules that remain persistent during the growing season (see figure 4b); (4) **spur shoot**: a short, sometimes sharp-pointed, slow-growing twig or branch covered with densely crowded leaf scars (see figure 4c); and (5) **thorn**: sharp-pointed modified twigs containing vascular tissue, and often bearing buds (see figure 4d and 4e).

Other outgrowths of the twig include **corky ridges, aerial roots,** and **tendrils.** Corky ridges are produced in some species, resulting in a "winged" twig (see figure 4f). This feature is particularly characteristic of winged elm (*Ulmus alata*) and sweetgum (*Liquidambar styraciflua*). Aerial roots and tendrils are characteristics of climbing woody vines (see figure 5h and 5i).

## Pubescence

Various types of plant hairs (trichomes) may be present on the buds, twigs, and leaves of woody plants. If hairs are completely lacking, the surfaces are referred to as **glabrous**. The terms "pubescent" or "hairy" are often used to describe the presence of a layer of detectable trichomes on the leaves, twigs, buds, or fruits. Hairs, if present, may be simple and unbranched or variously branched. **Stellate** hairs are starlike in shape. For additional listings of terms related to pubescence and surface coverings, see the **Glossary**.

## Bark

The bark of a tree can be divided into the outer bark (which is the dead layer composed of the cork) and the inner bark (which is the living layer composed primarily of the secondary phloem). This layer is separated from the wood (or secondary xylem) by the vascular cambium. It is the secondary phloem that transports food produced in the leaves to the remainder of the plant, and it is the secondary xylem that carries the water and nutrients from the roots to the remainder of the plant. The vascular cambium serves to continuously produce secondary phloem to the outside and secondary xylem to the inside during the growing season, and it is the difference in size between the spring wood and summer wood that results in growth rings. Each woody species tends to produce a particular type of bark (see following descriptions), and the bark features may provide useful features for identification. Bark also tends to vary with position and age, and the bark of branches may vary considerably from the bark of the trunk, and the bark of saplings and young trees may vary dramatically from the bark of mature specimens. In addition, some species may share similar bark features with other species, but there are usually subtle differences, and these can be distinguished with practice. Bark features described in this text, unless otherwise noted, are in reference to the mature trunk and not to the younger limbs and branches or saplings.

The following general types of barks (with examples) can be distinguished: smooth (*Carpinus*), scaly (*Ostrya*), shaggy (some *Carya*), platy (*Platanus*), blocky (*Diospyros*), papery (some *Betula*), furrowed (*Liriodendron*), smooth with vertical horizontal lines (*Prunus*), smooth with vertical lines (*Acer pensylvanicum*), exfoliating (*Viburnum molle*), peeling (*Betula alleghaniensis*), and shredding (*Lonicera japonica*). These latter four bark types are illustrated in Section V.

## CONES AND FRUITS

Woody plants produce seeds either in cones (if a gymnosperm) or in fruits (if an angiosperm). Most gymnosperms in Kentucky and Tennessee produce globose to ovoid scaly cones, these bearing the seeds attached to the thin, woody, or fleshy scales. Fruits are generally divided into those that are fleshy and those that are dry, and can be extremely valuable for identification, whether still attached to twigs or as withered remnants on the ground. It is often possible to determine whether the fruits are derived from inferior or superior ovaries (detectable by the position of the calyx remnants) for some fruit types, especially for capsules and for the several types of fleshy fruits, and this determination can be helpful in identification (see discussion under Berry in the **Glossary**). For obvious reasons, fleshy fruits are susceptible to winter freeze and easily deteriorate. Also, birds and other animals often favor fleshy fruits, especially berries and drupes. Dry fruits are more likely to persist in winter, and capsules are the fruit type most frequently encountered in numerous and often unrelated genera. For fruit and cone types see figures 6, 7 and 8, and for definitions of these fruit types see the **Glossary**.

# GYMNOSPERMS VS. ANGIOSPERMS

The two major categories of woody plants in Kentucky and Tennessee are the gymnosperms and the angiosperms. *Gymnosperm* is derived from Latin and means "naked seed," and the term *angiosperm* translates as "covered seed." These terms refer to the manner in which seeds develop, either naked and exposed on cone scales, as in gymnosperms, or borne completely enclosed in an ovary, as in angiosperms. The ovary is defined as the basal, ovule-bearing portion of the pistil, and in a typical flower the pistil is surrounded by the pollen-bearing stamens and the sterile perianth parts (sepals and petals).

Gymnosperms are a much more ancient group than the angiosperms, dating back to over 300 million years ago, and were the dominant plants during most of the dinosaur era beginning about 250 million years ago. Angiosperms first appeared about 140 million years ago, had became a significant part of the world's flora by 65 million years ago, and have continued to expand their diversity and range since this time period. Today angiosperms are by far the dominant type of terrestrial plants on the Earth from both an ecological and an economic standpoint.

Currently it is estimated that about 300,000 species of angiosperms exist on the Earth, compared to only about 1,000 species of gymnosperms. Angiosperms are the characteristic plants that dominate most of our major terrestrial biomes and ecosystems, such as grasslands, temperate deciduous forests, and tropical evergreen forests. Gymnosperms are the dominate plants in only a few types of biomes and ecosystems, but these often cover large regions of the Earth's surface, and are of great ecological and economic importance. Examples of these communities include the great conifer forests that occur in northerly latitudes and high-elevation regions around the globe, as well as the extensive pine forests of the southeastern United States.

Here are some notable facts about KY–TN gymnosperms: (1) the only deciduous-leaved species is the bald cypress (*Taxodium distichum*); (2) the only opposite-leaved ones are species of *Juniperus* and *Thuja*; (3) populations of *Thuja occidentalis* and *Taxus canadensis* are rare in the region, disjunct from larger populations to the east and north, and can be considered as relics of the Ice Age climates during the glacial periods; (4) native species of *Abies* and *Picea* are absent from Kentucky but there is a single species of each genus present in Tennessee and the two species form a distinctive spruce-fir zone mostly above 1700 meters elevation in the Smoky Mountains, where both species have undergone great population declines in the last few decades, most likely as a result of the combined effects of air pollution and exotic insect pests; (5) red-cedar (*Juniperus virginiana*), although very common and weedy over much of the region, is also associated with cedar-glade habitats, which include some of the most endangered species in the two-state region; (6) all four pine species native to Kentucky can be viewed at Sky Bridge in the Red River Gorge; (7) white pine (*Pinus strobus*) is the only native species of soft pine, these characterized by needles in 5s, a single vascular bundle in the needles, and unarmed cone scales, in the two states; the other five species of *Pinus* covered in the text are hard pines, characterized by needles in 2s or 3s, two vascular bundles in the needles, and prickly cone scales; and (8) both species of native hemlock (*Tsuga canadensis* and *T. caroliniana*) are currently threatened by the hemlock woolly adelgid, and this pest is likely to have devastating effects on these species and their associated communities.

## VEGETATIVE DIFFERENCES BETWEEN GYMNOSPERMS AND ANGIOSPERMS

Gymnosperms and angiosperms can be easily distinguished by their internal and external vegetative features. All gymnosperm species are woody, either shrubs or trees, whereas angiosperms include herbaceous species (annuals, biennials, perennials) as well as woody vines, shrubs, and trees. Gymnosperms have more primitive and less efficient vascular systems, with the water-conducting cells (xylem tissue) in their wood consisting of narrow tracheid cells, and the food-conducting cells (phloem tissue) in their bark consisting of sieve cells and albuminous cells. Angiosperms are characterized by a more advanced vascular system, including large-diameter vessel cells for water conduction and more efficient sieve tube elements and companion cells for food conduction.

The gymnosperms in our region all produce narrow, flattened or angled, needlelike leaves, these few-veined (usually with only one or two veins), and the veins running parallel through the needle. In addition, all gymnosperms in our area are evergreen except for the genus *Taxodium*. Angiosperms produce a great variety of leaf types, but mostly with the blades wide and flat (rarely linear or needlelike) and with multiple veins (usually with a netlike or parallel pattern). Most woody angiosperms in our region have a deciduous habit, but there about 50 genera in which at least some of the species are evergreen.

The wood of both gymnosperms and woody angiosperms is produced in annual growth rings, but the rings tend to be much more pronounced in angiosperms because of the greater size difference between summer wood cells and spring wood cells. Gymnosperm wood is considered to be "nonporous" as it lacks the vessel cells that are characteristic of angiosperms. Angiosperm wood is considered to be "porous," as the large vessels are readily evident and appear to form rings of large pores when viewed in cross-section. Angiosperm wood can be further divided into the "ring-porous" wood and "diffuse-porous" wood. Ring-porous wood can be found in species of oak, ash, hickory, and elm, and is characterized by annual growth rings in which the spring wood cells are conspicuously larger than the summer wood cells. Tuliptree, as well as species of birch and maple, produce annual rings in which there is not a distinct seasonal variation in size, but a more gradual change from the larger spring wood cells to the smaller summer wood cells.

## Reproductive Differences

Gymnosperms typically require a prolonged period for seed production, often a year or more, whereas angiosperms are generally much more rapid, usually with only a few weeks or months between pollination and seed production. Gymnosperms are largely wind-pollinated, whereas the majority of angiosperms are insect-pollinated (but some, such as oaks and hickories, are wind-pollinated). There are also many other differences in their life cycles, including the method by which sperm is delivered to the egg, the fewer numbers of cells around the egg and in the pollen grain of angiosperms, and in the method of embryo development. True fruits are found only in angiosperms, although the seed-bearing cones of gymnosperms have sometimes been referred to as fruits in the literature. Fruits are therefore unique to angiosperms, formed by the maturing ovaries, and function as an advanced form of seed dispersal. There is a tremendous variety of fruit types, varying from fleshy or nutlike fruits that are eaten by animals, to winged fruits that are dispersed by the wind, to dry fruits with hooks or barbs that attach to passing animals (or people). In many cases even the seeds have modifications to aid in wind or animal dispersal. In gymnosperms the seeds are dispersed by wind or gravity or in a few cases by animals (as in *Taxus*).

## Classification

All gymnosperms of Kentucky and Tennessee are conifers, classified in the Division Coniferophyta (or Pinophyta). Other major groupings of gymnosperms include the cycads and the gingkos, which have fernlike or fanlike leaves, respectively, and differ from conifers in a number of ways relating to their reproduction. The Asian ginkgo (*Ginkgo biloba*) can be readily cultivated in the region but does not become naturalized; cycads will not grow in our region because they require tropical or subtropical climates. Within the southeastern United States, cycads (*Zamia* spp.) are found in only Florida and southern Georgia.

Angiosperms have traditionally been divided into two classes: Monocotyledonae (monocots), with one cotyledon (seed leaf) in their embryos, and Dicotyledonae (dicots), with two cotyledons. With recent advances in genetic studies this classification is now considered outdated, as there are several lineages of traditional "dicots" (e.g., families in this treatment such as Annonaceae, Calycanthaceae, Lauraceae, and Magnoliaceae) that are now better understood as basal "primitive" angiosperms not closely related to the core dicots. For the purposes of this text, however, it is useful to point out that woody monocots can easily be distinguished from woody "dicots" by the following features: (1) vascular bundles in the stems of monocots are in a scattered arrangement, but are in a cylindrical arrangement in the stems of dicots; (2) the leaf venation is parallel or arcuate (lateral veins arching toward leaf apex) in monocots but is typically either net pinnate or net palmate in dicots; and (3) flower parts (sepals and petals) are in 3s or in highly modified spikelets in monocots, but are usually in 4s or 5s in dicots (this numerical feature is sometimes evident in the remnant fruits). There are only a few woody monocots in the KY–TN flora, including two species of *Arundinaria*, six species of *Smilax*, and two species of *Yucca*. The other woody angiosperms covered in this book have traditionally been classified as dicots.

## Rare Woody Plants

Based on the listings maintained by the Kentucky State Nature Preserves Commission (KSNPC 2010) and by the Tennessee Natural Heritage Program (Crabtree 2012), there are 90 rare woody plant species in Kentucky and Tennessee. Only two woody species in KY–TN are listed at the federal level—*Conradina verticillata* and *Spiraea virginiana*, and these species occur in both states. Of the rare woody plants known from 6 or fewer counties, 5 are known from KY only, 35 from TN only, and 7 from KY and TN, with the listing as follows: **KY only**: *Juniperus* *communis, Malus ionensis, Salix amygdaloides, Salix discolor, Symphoricarpos albus*; **TN only**: *Abies fraseri, Alnus viridis* var. *crispa, Betula cordifolia, Buckleya distichophylla, Clethra alnifolia, Cotinus obovatus, Crataegus harbinsonii, Diervilla rivularis, Fothergilla major, Gaylussacia dumosa, Gelsemium sempervirens, Kalmia carolina, Leucothoe racemosa, Linnaea borealis, Lonicera canadensis, Lonicera flava, Magnolia virginiana, Menziesia pilosa, Nemopanthus collinus, Neviusia alabamensis, Pieris floribunda, Polygonella americana, Prunus pumila, Quercus margaretta, Rhamnus alnifolia, Ribes aureum, Ribes curvatum, Sibbaldiopsis tridentata, Smilax laurifolia, Symplocos tinctoria, Tsuga caroliniana, Ulmus crassifolia, Vaccinium elliottii, Vaccinium macrocarpon*, and *Viburnum bracteatum*; and **KY and TN**: *Ceanothus herbaceus, Comptonia peregrina, Conradina verticillata, Nestronia umbellula, Ribes americanum, Schisandra glabra*, and *Spiraea alba*.

## NON-NATIVE WOODY PLANTS

Currently about 17% of the woody flora of KY–TN is non-native. Of the 76 non-native species, about 40 are listed by the state exotic plant pest councils (KY-EPPC 2011, TN-EPPC 2009) as posing serious ecological threats in the region. Some of these species are vinelike and have the ability to spread vegetatively and overgrow native vegetation or form extensive ground covers, such as *Akebia quinata, Celastrus orbiculatus, Euonymus fortunei, Clematis terniflora, Hedera helix, Lonicera japonica, Pueraria montana, Vinca* spp., and *Wisteria* spp. Many non-native species are shrubby, and these often cause problems by occupying the understory in forested lands and inhibiting the growth of native herbs, shrubs, and small trees, as well as the seedlings and saplings of the canopy species; examples of these shrubby understory exotics include the following: *Berberis thunbergii, Elaeagnus* spp. (especially *E. umbellata*), *Euonymus* spp. (especially *E. alatus*), *Ligustrum* spp., *Lonicera* spp. (especially *L. maackii* and *L. fragrantissima*), *Rhamnus* spp., and *Rosa* spp. (especially *R. multiflora*). There are only a few species of non-native trees, including *Ailanthus altissima, Albizia julibrissin, Broussonetia papyrifera, Koelreuteria paniculata, Maclura pomifera, Melia azedarach, Morus alba, Paulownia tomentosa, Populus alba*, and *Pyrus calleryana*. These non-native trees tend to occur in disturbed areas and along forest borders. Another set of non-native species are those that are not yet widespread in KY–TN but are likely to become more troublesome in the near future, including such species as *Mahonia bealei, Buddleja davidii, Nandina domestica*, and *Rubus bifrons*.

# USES OF WOODY PLANTS FOR FOOD, MEDICINE, FIBER, AND WEAPONS

In the generic and species accounts there are many instances where notes are included on the uses of the particular genus or species for food, medicine, fiber, and weapons. In addition, there are four appendixes (II to V) that provide lists of the most readily available plants for the above-listed purposes. This information is provided only for educational purposes. We realize that many people have interests in the uses of products available in nature, and it is our hope that these kinds of information will provide a more enriching reading experience, and perhaps stimulate the reader to learn more. It is not our intent to advise the reader on the best treatment for any particular ailment or injury or in any particular emergency situation where medicines, fiber, or weapons might be needed. Nor is it our intent to instruct the reader on the particulars of proper preparation of foods, medicines, fiber, and weapons. It is our goal to provide the reader with a means of proper identification of species that might have potential uses in the winter, and it is up to the readers to seek other books and references for additional information (see Sources of Information and Literature Cited) to help identify the best methods of preparation for these resources and to use their own best judgment in the uses of these products. Especially recommended are the books by Burrows and Tyrl (2001) and Moerman (1998), which provide encyclopedic coverage of toxic plants and Native American ethnobotany, respectively.

Those interested in preparing for emergency situations, where they might encounter plants of unknown edibility, should become familiar with the **Universal Edibility Test** (available at many websites). We do strongly advise those interested in exploring the uses of wild plants to learn as much as possible about the families, genera, and species of wild plants in the region, including proper identification techniques. Enroll in college classes, short courses, or workshops that address these topics. Some plants harbor deadly poisons, others may cause temporary illness or hallucinations, and others are completely harmless but may or may not be a good food source. Many species are edible only at certain times of the year or after baking or boiling or straining. There are a number of cases where an edible species has a poisonous look-alike. As a general rule, no plant should be eaten unless a positive identification can be made. See **Appendix II** for a list of readily available food plants in winter. General categories and conditions of plants to avoid include:

- Plants with white fleshy fruits (berries or drupes)
- Plants with milky or colored sap
- Plants with three leaflets on a long stalk
- Plants with bean-like pods
- Plants with swollen below-ground bulbs
- Plants with a bitter or soapy taste
- Plants with dense hairs, spines, or thorns
- Plants with an almondlike scent (cherries and relatives)
- Plants with much-divided, parsleylike foliage and flat-topped (umbels) clusters of flowers (carrot family)
- Plants similar to tobacco or potato (Solanaceae family)
- Plants growing in potentially contaminated soils or water
- Plants with wilted or fungal-infected parts

For medicinal uses of plants in winter, we have emphasized those plants that can be used for treating wounds and other external problems, assuming these kinds of medical problems would be the most common during winter emergencies (**Appendix III**). For other situations we strongly urge that self-medication with wild plants be avoided, and advice should be sought from a physician or other professional health-care practitioner.

Some sort of string or rope is often indispensable in emergency situations, and many woody plants (and some herbaceous ones) provide a ready source of fiber for constructing various kinds of cordage (**Appendix IV**). As a general rule, the dried inner bark from dead tree trunks is the best sources of these fibers, and these can be twisted together to form relatively strong cordage. If a dead tree is not available, it may be necessary to use the fiber from living trees, and in this case it is best to use the fiber from young branches (or even whole slender branches can be used), but this requires the drying of the moist inner bark layers. If it is deemed necessary to remove the inner bark from the trunks of living trees for cordage (or for food or for preparing a poultice for wounds), be careful to remove the fibers in vertical strips, and avoid cutting around the trunk, which will girdle the tree and kill it. The preparation of cordage from the moist inner bark often requires weeks of soaking and drying.

**Appendix V** provides a list of genera that provide good materials for the construction of bows and arrows. Rough bows and arrows can be made in the field, and some local and abundant plants (osage-orange, wild cane) provide excellent raw materials. Yews (*Taxus* spp.) have been a preferred wood for bows for thousands of years, but native yews in our region are now scarce, and should not be exploited for this reason. The construction of bows and arrows can also be a fine art, requiring extreme attention to wood selection and shaping, notching, and stringing, and there are detailed instructions at many websites and in books for those interested in practicing this craft. Among the better books on wilderness survival skills are the series of books written by John and Geri McPherson, as well as the many books authored or coauthored by Tom Brown Jr.

## Conservation Concerns

As a general rule we advise those seeking wild plants for food or medicines to select non-native species, if the choice is available. In particular, learn to identify those native plants that are rare or endangered and avoid harming these populations unless there is a life-threatening situation. Many non-native plants with edible parts are readily available in many parts of our region, such as species of *Elaeagnus* and *Rosa*, as well as a number of exotic herbs (**Appendix II**). If using native plants, first determine the size of the population, and try to leave sufficient numbers of plants to restore the population, and avoid taking any more than is absolutely necessary for your purposes.

## Disclaimer

We must emphasize that the information related to the use of wild plants for foods and medicines in this book is provided only as a general guide. Some people may experience harmful reactions to wild plant foods and medicines that are generally regarded as safe for most people. In particular, the information in this book should not be used for self-medication without consulting a physician—this can be very dangerous. Attempts to utilize wild plants in any way depend on many factors controllable only by the reader, and the authors and publisher assume no responsibility in the case of adverse effects in individual cases.

# Section II. Key to Keys

1. Woody plants EITHER with succulent stems or leaves (cactus or yucca) OR with grasslike habit (wild cane with parallel-veined leaves and hollow stems) _____**Key A, p. 23.**
1. Woody plants lacking succulent parts and grasslike habit (woody vine, subshrub, shrub, or tree).
    2. Gymnosperms; mostly trees (shrubby in species of *Taxus* and *Juniperus*); leaves needlelike or scalelike, arrangement various, but if clustered then needles 2–5 in number; leaves evergreen in all genera except *Taxodium* (note fallen twigs with linear leaves, and branches with globose seed cones and/or pendulous panicles of immature pollen cones, these with pinelike odor); seeds borne naked, in cones (except in *Taxus*) _____**Key B, p. 23.**
    2. Angiosperms; trees, shrubs, or woody vines; leaves evergreen or deciduous, typically with a flat blade and not needlelike (if somewhat needlelike then plants shrubby, and EITHER with leaves in clusters of 6 or more and minty aromatic or punctate, as in *Conradina* and *Hypericum*, OR stems jointed with nodal rings, as in *Polygonella*); seeds enclosed in a fruit, the remnant fruits sometimes present in winter.
        3. Leaves evergreen, typically with thick texture and green color.
            4. Leaves opposite or whorled _____**Key C, p. 24.**
            4. Leaves alternate _____**Key D, p. 25.**
        3. Leaves deciduous, absent in winter (some species with persistent, tardily deciduous, remnant leaves, these turning brown).
            5. Plants with armature (thorns, spines, or prickles) _____**Key E, p. 27.**
            5. Plants lacking armature.
                6. Stems viny, requiring support, either climbing by tendrils, aerial roots, or by twining, or trailing and usually rooting at nodes; plants in this category do not produce erect stems > 30 cm tall that are capable of supporting their own weight (some *Vitis* species may be shrublike and lack tendrils, but have shreddy bark and brown pith with a partition at the node, unlike any species in following categories) _____**Key F, p. 29.**
                6. Stems erect, either trees or shrubs, or if sprawling or trailing, then producing erect stems > 30 cm tall, only rarely rooting at the nodes, as in *Decodon*.
                    7. Leaf scars opposite or whorled.
                        8. Leaf scars of young twigs joined around the twig or connected by lines (connecting lines may be obscured in older twigs) _____**Key G, p. 31.**
                        8. Leaf scars of young twigs not joined or connected by lines around the twig _____**Key H, p. 33.**
                    7. Leaf scars alternate.
                        9. Catkins or densely flowered racemes present in winter _____**Key I, p. 34.**
                        9. Catkin-like structures absent in winter.
                            10. Twigs with encircling stipular scars, these > ¾ encircling the twigs at each node, OR with leaf scars > ½ encircling the twigs _____**Key J, p. 35.**
                            10. Twigs lacking encircling stipular scars, or if present then < ¾ encircling the twigs, and leaf scars < ½ encircling the twig.
                                11. Pith not solid and continuous, either chambered or diaphragmed or with spongy cavities or hollow_____**Key K, p. 35.**
                                11. Pith solid and continuous, lacking cavities or partitions.
                                    12. Bundle scar clearly 1 (obscure in *Cytisus*, with twigs green, angled, and flexible) _____**KEY L, p. 36.**
                                    12. Bundle scars 3 or more (if obscure, then twigs not as described above).
                                        13. Twigs with true end buds OR end buds clustered _____**KEY M, p. 37.**
                                        13. Twigs lacking true end buds, the false end buds not clustered (lateral buds sometimes clustered or superposed) _____**KEY N, p. 40.**

# Section III. Keys to Genera

## KEY A
Plants succulent or grasslike (a cactus, wild cane, or yucca).

1. Stems green, flattened, and padlike, bearing clusters of sharp needlelike spines and tiny barbed bristles; a cactus _____**Opuntia.**
1. Stems otherwise, either slender and hollow, or short and thick, lacking spines and bristles as described above.
   2. Leaves thin and grasslike (may be absent in winter), to 3 dm long, and lacking threadlike filaments along the margins, the leaves scattered along the stem; stem slender and hollow, to 3 m (rarely 8 m) tall; fruit a grain, in spikelets; a wild cane _____**Arundinaria.**
   2. Leaves fleshy, sword-like, and spine-tipped, to 8 dm long, with threadlike filaments along the margins, the leaves crowded at or near ground level, and concealing the short ($< 0.5$ m) thick trunk; fruit a dry dehiscent capsule, borne in panicles on a bracteate stalk emerging from the center of the leaf cluster, and reaching 3 m tall; a Spanish bayonet_____**Yucca.**

## KEY B
Gymnosperms with cones and/or naked seeds; leaves needlelike or scalelike.

1. Leaves absent in winter (remnant or fallen twigs with leaves linear and flat, about 1 mm wide and 1.5 cm long, in one plane); twigs with branch scars, knobby buds, tiny stiff scale leaves, globose cones, and pendulous panicles of immature pollen cones, these with pinelike odor _____ **Taxodium.**
1. Leaves present in winter; twigs, cones, and pollen cone clusters otherwise.
   2. Leaves opposite or whorled; the shape scalelike (or sometimes needlelike in juvenile forms).
      3. Branchlets flattened; cones scaly and brown_____**Thuja.**
      3. Branchlets rounded; cones fleshy and blue _____**Juniperus.**
   2. Leaves alternate, spiraling, or in clusters of 2–5, needlelike and angled or flattened.
      4. Needles in clusters of 2, 3, or 5, angled, the bases enclosed in a sheath (sometimes deciduous); cones with woody scales, these prickly (except in *P. strobus*) _____**Pinus.**
      4. Needles attached singly, angled or flattened, the bases not enclosed in a sheath; cones, if present, lacking prickles.
         5. Shrubs with green twigs; needles flat and sharp-pointed, 10–20 mm long, narrowed to a short petiole that continues as a narrow ridge on the twig; scaly cones absent, the seeds surrounded by red, fleshy aril _____**Taxus.**
         5. Trees with brownish or grayish twigs; needle shape and attachment otherwise; scaly cones present.
            6. Needles angled in cross-section, sharply pointed, dark green, about 15 mm long, attached directly to stout woody pegs and needles lacking petioles; cones about 4 cm long, the scales entire or nearly so_____ **Picea.**
            6. Needles flat, the size various, blunt-tipped, pale or whitened below, either petiolate or narrowed to base; cones various.
               7. Needles 8–15 mm long, with distinct petiole attached to a tiny woody peg (the leaf scars therefore elevated on the peg); cones about 4 cm long, with thin, rounded scales___**Tsuga.**
               7. Needles 12–25 mm long, narrowed to a disc-like base on twig, the leaf scars rounded and flush with twig surfaces; cones to 10 cm long, but cone scales falling from axis in winter, the spike-like cone axis sometimes persistent _____**Abies.**

# KEY C

**Angiosperms with leaves evergreen and opposite or whorled.** Note: *Buddleja davidii*, with leaves persistent in leaf axils, and white-woolly on lower surfaces, is included in Key G.

1. Plant a shrubby parasite attached to tree branches, the twigs green; berries whitish and sticky, often present in winter _____**Phoradendron.**
1. Plants otherwise, rooted in soil.
   2. Plants with shrubby habit, typically with upright branches, or sometimes sprawling, but usually with aerial branches > 20 cm tall (including *Conradina*, with minty odor); plants in this category not herblike or viny (*Vinca major*, with ciliate leaf margins and forming a ground cover, is keyed under next couplet).
      3. Leaves serrate.
         4. Plant a low shrub, spreading by stolons, the aerial stems to 4 dm tall; bundle scar 1; leaf blades < 1 cm wide, with smooth surfaces; plants to 4 dm tall; rare native species associated with calcareous soils_____ **Paxistima.**
         4. Plant a shrub > 10 dm tall; bundle scars 3 (if bundle scar 1, see note on evergreen shrubs under *Euonymus* in Section IV); leaves > 2 cm wide, with wrinkly-rugose veins; non-native species of disturbed areas (*V. rhytidophyllum*)_____**Viburnum.**
      3. Leaves entire.
         5. Leaves linear or linear-lanceolate, in axillary clusters; leaves either aromatic or with punctate dots.
            6. Plants sprawling, and rooting at nodes, with branches < 3 dm tall; twigs and foliage with minty odor; fruit of 4 nutlets; rare species of restricted distribution, chiefly associated with gravelly or sandy stream banks of Big South Fork region of KY and TN _____ **Conradina.**
            6. Plants usually upright, some species reaching 2 m tall; twigs lacking minty odor, but leaves with punctate dots; fruit a capsule; distributions more widespread_____**Hypericum.**
         5. Leaves wider, lanceolate to elliptic to ovate, not in axillary clusters; leaves neither aromatic nor with transparent dots.
            7. Shrubs to 1.5 m tall; leaves thick and shiny-leathery, clustered toward twig tips; fruit a capsule, often persistent; rare and local native species restricted to E TN.
               8. Leaves oblong, to 6 cm long, opposite or whorled (*K. carolina*) _____**Kalmia.**
               8. Leaves to 2.5 cm long and 8 mm wide, opposite or alternate _____ **Leiophyllum.**
            7. Shrubs to 3 m tall; leaves various, but not clustered toward twig tips; fruit fleshy; non-native species of disturbed sites.
               9. Buds naked; leaf surfaces with wrinkly-rugose veins; twigs densely tomentose (*V. rhytidophyllum*) _____**Viburnum.**
               9. Buds scaly; leaf surfaces relatively smooth; twigs smooth or slightly hairy.
                  10. Bundle scar 1; drupes blackish and often persistent in terminal panicles____ **Ligustrum.**
                  10. Bundle scars 3 or obscure; berries reddish, in pairs in leaf axils, but usually absent in winter _____ **Lonicera.**
   2. Plants EITHER herblike with stems < 20 cm tall OR with trailing or viny habit (excluding species with minty odor but including species with milky sap).
      11. Leaves simple, to 2 cm long, about as wide as long; plants creeping, native to the Appalachians.
         12. Leaves entire _____ **Mitchella.**
         12. Leaves serrate _____**Linnaea.**
      11. Leaves simple or compound, the leaves or leaflets > 2 cm long, usually longer than wide (except sometimes in *Vinca major*); plants of various habits and distribution, including non-native species.
         13. Plants herblike, with erect stems to 20 cm tall; leaves sharply toothed, usually with white color along veins _____ **Chimaphila.**
         13. Plants either creeping and forming a ground cover or climbing; leaves various.
            14. Leaves compound.
               15. Leaflets 2; plants climbing by tendrils (modified leaflet) _____**Bignonia.**
               15. Leaflets 5; plants climbing by clasping petioles (*C. terniflora*)_____ **Clematis.**

14. Leaves simple.
    16. Plants not climbing, with short upright stems and spreading stolons, often forming a dense ground cover; leaves entire, 2–6 cm long; milky sap sometimes present; non-native _____ **Vinca.**
    16. Plants climbing (or sometimes forming a ground cover, as in *Euonymus fortunei* with serrate leaves); leaves often longer; milky sap absent.
        17. Leaves serrate; plants climbing or spreading by aerial roots.
            18. Twigs green, often angled, bundle scar 1; non-native vines with glossy-green leaves (*E. fortunei*) _____ **Euonymus.**
            18. Twigs brown, terete, but with a jointed or zigzag appearance, the nodes swollen; bundle scars 3; native vine with tardily deciduous leaves_____ **Decumaria.**
        17. Leaves entire (sometimes ciliate), or sometimes with a few lobes (as in *Lonicera*); plants climbing by twining, twigs brown or purplish.
            19. Twigs pubescent; fruit a blackish berry in pairs in leaf axils; widespread exotic species (*L. japonica*) _____ **Lonicera.**
            19. Twigs smooth; fruit otherwise; native species, the distributions various.
                20. Bundle traces 3; uppermost leaves surrounding the stems (connate-perfoliate); fruit a reddish berry; species widely distributed_____ **Lonicera.**
                20. Bundle scar 1; uppermost leaves attached to side of stem; fruit a dry capsule; species restricted to southeastern TN _____ **Gelsemium.**

# KEY D
**Angiosperms with leaves evergreen and alternate.**

1. Leaves compound (if a small tree with green twigs and flat-based spines, and leaflets 3, see *Poncirus* in Key E).
    2. Plants herblike, low-growing, unarmed, with stems < 10 cm tall, arising from creeping underground rhizomes; leaflets 3; native species of high elevations in E TN _____ **Sibbaldiopsis.**
    2. Plants obviously woody, prickly in some, with stems much taller or longer; leaflet number variable; native or non-native at various elevations.
        3. Stems or leaves prickly or bristly; native and non-native species.
            4. Leaves pinnately compound with > 5 leaflets, these leathery and spiny-margined, similar to the leaves of *Ilex opaca* (American holly)_____**Mahonia.**
            4. Leaves with 3 or 5 leaflets, these membranous, not similar to American holly.
                5. Petiole incompletely deciduous, the basal portion remaining as a "petiolar stump"; leaf scars hidden under petiole base _____ **Rubus.**
                5. Petiole completely deciduous; leaf scars evident and forming a narrow line partially encircling the twig _____**Rosa.**
        3. Plants unarmed; non-native species.
            6. Plant a vine; leaves palmately compound_____ **Akebia.**
            6. Plant a shrub; leaves bipinnately to tripinnately compound _____ **Nandina.**
1. Leaves simple (if pinnately lobed and with aromatic odor, see *Comptonia* in Key I).
    7. Plants climbing by tendrils or aerial roots.
        8. Plants climbing by tendrils; major secondary veins green and arching toward apex; stems often prickly_____**Smilax.**
        8. Plants climbing by aerial roots; major secondary veins whitish and either palmately or pinnately arranged; stems lacking prickles_____ **Hedera.**
    7. Plants trailing to erect.
        9. Plants herblike, barely woody, < 20 cm tall.
            10. Leaves sharply toothed, the veins whitish _____ **Chimaphila.**
            10. Leaves entire to slightly crenate-serrate, the veins greenish.

11. Leaves cordate at base, lacking aromatic odor _____ **Epigaea.**
11. Leaves tapered or rounded to base, with wintergreen odor when crushed _____ **Gaultheria.**
9. Plants obviously woody, usually erect and > 20 cm tall, or if sprawling and low-growing, then foliage not as described for three genera above.
    12. Leaves palmately lobed (at least some), and veins whitish; plants trailing and forming a ground cover, the stems producing numerous adventitious roots _____ **Hedera.**
    12. Leaves otherwise, varying from entire, to few-toothed, to serrate or spiny-margined; erect shrubs or trees, or if trailing, then leaves not lobed.
        13. Shrubs < 1 m tall, with jointed stems and nodal sheaths encircling stems; leaves linear, about 1 mm wide and thick, to 15 mm long _____ **Polygonella.**
        13. Shrubs or trees, most species > 1 m tall (if < 1 m tall, then leaves much wider).
            14. Leaves dotted below with punctate glands or scales; all shrubs.
                15. Leave dull green above, thin but persistent, varying from entire to few toothed, with punctate glands; fruit a cypsela with tufts of white bristles, in heads____ **Baccharis.**
                15. Leaves glossy green above, thick and obviously evergreen, entire or inconspicuously serrulate, with dots or scales, generally elliptic; fruit otherwise.
                    16. Leaves punctate with glandular dots below, the lower surface pale green otherwise; fruit a capsule; native species of mid- to high elevations in TN (*R. minus*) _____ **Rhododendron.**
                    16. Leaves dotted with reddish and silvery scales below; fruits fleshy and reddish with brown scales; non-native species _____ **Elaeagnus.**
            14. Leaves lacking punctate glands or scaly dots on lower surface (stalked glandular trichomes present in *Pieris*); subshrubs, shrubs, or trees.
                17. Pith chambered; leaves elliptic to oblanceolate, entire or minutely toothed, about 10 cm long and 3 cm wide, the leaves sweet-tasting; fruit fleshy and single-seeded; small tree or shrub _____ **Symplocos.**
                17. Pith continuous or diaphragmed; leaves various, but not sweet-tasting; fruit dry or fleshy and several-seeded.
                    18. Trees often reaching > 20 m tall; twigs with encircling stipular scars; leaves to 25 cm long, entire; bundle scars many _____ **Magnolia.**
                    18. Shrubs (tree heights of 10–15 m only in *Rhododendron* and *Ilex*); twigs lacking encircling stipular scars; leaves various; bundle scar one.
                        19. Leaves spiny-margined or serrate.
                            20. Leaves with sharp spines along margins, glossy green; stipular scars or tiny black stipules present; fruit fleshy and berrylike with < 10 seeds; trees to 15 m _____**Ilex.**
                            20. Leaves crenate to sharply serrate, lacking spines; stipules absent; fruit a capsule or a 10-seeded berry; shrubs < 2 m tall.
                                21. Shrubs to 30 cm tall; leaves to 2.5 cm long; plants colonial from long rhizomes, forming a dense ground cover; fruit a 10-seeded berry; plants rare and local (*G. brachycera*) _____ **Gaylussacia.**
                                21. Shrubs to 2 m tall; leaves 3–13 cm long, plants not forming a dense ground cover (but sometimes forming tangled thickets); fruit a capsule.
                                    22. Twigs with coarse stiff hairs; leaf surfaces with glandular trichomes; leaf margins crenate-ciliate; capsules globose _____ **Pieris.**
                                    22. Twigs softly pubescent or glabrous; leaf surfaces eglandular; leaf margins serrate; capsules wider than long _____ **Leucothoe.**

19. Leaf margins entire.
    23. Leaves < 8 mm wide; shrubby species to 2 m tall, highly restricted in distribution in TN.
        24. Leaves very crowded, with mix of alternate and opposite leaves; erect shrub to 2 m tall; fruit dry; plants of dry uplands and balds at mid to high elevations in TN _____ **Leiophyllum.**
        24. Leaves regularly spaced, clearly alternate; low trailing shrub to 1 m tall; fruit fleshy; rare wetland plants (*V. macrocarpon*) _____ **Vaccinium.**
    23. Leaves > 10 mm wide; shrubs or small trees typically > 2 m tall; species of more general distribution in TN and KY.
        25. Leaves 2–5 cm long, less than twice as long as wide, not clustered toward apex; buds globose, < 5 mm long; terminal buds absent (*V. arboreum*) _____ **Vaccinium.**
        25. Leaves 5–20 cm long, more than twice as long as wide, clustered toward apex; buds ovoid, 10–30 mm long; terminal buds present.
            26. Petiole bases of older leaves concealing the axillary buds; leaf apices acute to acuminate; capsules depressed-globose (*K. latifolia*) _____ **Kalmia.**
            26. Petiole bases of older leaves not concealing axillary buds; leaf apices rounded to short-acute; capsules cylindric _____ **Rhododendron.**

# KEY E
**Angiosperms with leaves deciduous; plants with armature (thorns, spines, or prickles).**

1. Leaf scars opposite or subopposite; twigs thorn-tipped.
   2. Bundle scars 3; buds rounded, usually solitary at nodes and appressed against the twig; bud scales dark brown with white-ciliate margins _____ **Rhamnus.**
   2. Bundle scar 1; buds acute-tipped and divergent from stem, often superposed; bud scales brown or gray, keeled, lacking ciliate margins _____ **Forestiera.**
1. Leaf scars alternate; armature various.
   3. Stems climbing by tendrils (modified stipules!) emerging from persistent leaf bases; armature of epidermal prickles, usually scattered, sometimes scarce; vascular bundles scattered and pith absent (monocots) _____ **Smilax.**
   3. Stems otherwise, if climbing, then lacking tendrils; vascular bundles in a cylinder and pith present (dicots).
      4. Armature all or chiefly of paired unbranched nodal spines or prickles; internodal spines or prickles usually absent (twigs may be bristly).
         5. Shrubs or vines; armature of prickles, these often with extended broadened bases, and easily snapped off; leaf scars very narrow and about ½ encircling the stem; buds small, clearly few-scaled, often reddish; fleshy red fruits (hips) often persistent _____ **Rosa.**
         5. Small to large trees; armature of woody modified stipules, not easily removed; leaf scars, buds, and fruits not in above combination.
            6. Buds evident and red-hairy, located above the leaf scar; terminal bud present; spines situated just below level of leaf scar; twigs with citrus aroma _____ **Zanthoxylum.**
            6. Buds obscure, sunken in leaf scar; terminal bud absent; spines directly adjacent to leaf scars; twigs lack citrus aroma; fruit a legume, often persistent _____ **Robinia.**
      4. Armature otherwise, of solitary or numerous spines at nodes, or chiefly of internodal spines or prickles, or of thorns (modified branches); these spines or thorns not regularly in pairs at the nodes, but sometimes branched.

7. Twigs very thick, 1–3 cm in diameter; leaf scars nearly encircling the twig; bundle scars > 20; spines scattered along twig surfaces; plant a small tree with spiny trunk _____ **Aralia.**
7. Twigs lacking above combination of features.
    8. Armature of epidermal prickles (and bristles in some), these often broad-based and flattened, and hooked or recurved; scattered along internodes and at nodes; plants often sprawling or trailing or vinelike.
        9. Petiole incompletely deciduous, the basal portion remaining as a "petiolar stump"; leaf scars hidden under petiole base_____ **Rubus.**
        9. Petiole completely deciduous; leaf scars evident and forming a narrow line partially encircling the twig_____**Rosa.**
    8. Armature of distinct spines or thorns, these often associated with the nodes or twig tips; upright shrubs or trees.
        10. Thorns to 20 cm long, frequently branched, on both twigs and trunks, and exceedingly sharp-pointed; twigs zigzag; buds sunken in bark, often superposed; end buds false; trees, often > 20 m tall _____**Gleditsia.**
        10. Thorns or spines much smaller; twigs and buds various, but buds usually emergent and evident; shrubs and small trees, usually < 15 m tall.
            11. Bundle scar 1 (plants EITHER with silvery scales, OR twigs green and flattened, OR vine-like habit); all exotic species.
                12. Twigs with silvery or brownish scales; fruit fleshy, yellowish, and scaly, sometimes persistent_____ **Elaeagnus.**
                12. Twigs lacking silvery or brownish scales; fruits otherwise.
                    13. Twigs and spines green, the spines flattened at base; small tree ____ **Poncirus.**
                    13. Twigs and spines grayish or brownish; spines rounded at base; sprawling shrub with poisonous red berries _____ **Lycium.**
            11. Bundle scars > 1 or obscure (plant features not as described above); native and exotic species.
                14. Spines attached below the leaf scar; low shrubs to 2 m tall; bark shredding or with loose plates in older stems; fruit a berry.
                    15. Spines attached below cluster of persistent petiolar stalks and buds; inner bark yellow; pith continuous_____ **Berberis.**
                    15. Spines attached at base of leaf scar; inner bark not yellow; pith spongy in older twigs _____ **Ribes.**
                14. Spines or thorns terminating twigs, or if axillary, then located above leaf scar or to one side; habit various, shrubs to small trees; bark various; fruit otherwise.
                    16. Shrubs to 3 m tall, with axillary thorns, these usually with a pair of buds at the base; exotic escape with applelike fruits_____ **Chaenomeles.**
                    16. Taller shrub or tree; thorns usually with a single bud at base, or buds lacking; native species, the fruits various, but if applelike, then some thorns in terminal position; fruit various.
                        17. Armature of highly modified, very sharp spines or thorns, these typically lacking buds; spines/thorns mostly lateral, in axillary position on the twigs.
                            18. Leaf scars narrow, crescent shaped; bundle scars clearly 3; end buds true; buds spherical and gummy; milky sap absent; fruit a pome, about 1.5 cm thick, sometimes persistent_____**Crataegus.**
                            18. Leaf scars ovoid or circular; bundle scars many or obscure; end buds various; spur branches often present; buds dome-shaped; milky sap present, especially in late winter; fruit otherwise.
                                19. True end buds present; inner bark red-brown; fruit elliptic, < 1.5 cm long, blackish and fleshy, with a single seed_____ **Sideroxylon.**

19. True end buds absent; inner bark yellow; fruit yellow-green, similar in size and shape to a grapefruit, with numerous seeds _____ **Maclura.**
17. Armature of sharply pointed twigs, these often with buds, and at least some in terminal positions on the twig.
   20. True end buds absent; twigs with almond odor; bark often smooth but with horizontal stripes (especially on younger stems and branches) _____**Prunus.**
   20. True end buds present; twigs lacking almond odor; bark scaly, lacking horizontal stripes.
      21. Buds dome-shaped, about as wide as long; milky sap sometimes present; fruit a drupe_____**Sideroxylon.**
      21. Buds ovate or lanceolate, longer than wide; milky sap absent; fruit a pome.
         22. Buds pubescent and blunt and to 7 mm long _____ **Malus.**
         22. Buds glabrous and acute, or pubescent and > 7 mm long _____**Pyrus.**

# KEY F
**Angiosperms with leaves deciduous; the plants unarmed and with viny or trailing habit.**

1. Leaf scars opposite or whorled.
   2. Plants climbing by aerial roots, lacking tendrils or twisting petioles.
      3. Leaf scars large and shield shaped, with 1 crescent-shaped bundle scar (or broken into several); terminal bud absent; capsules cylindric, to 15 cm long, with winged seed; common species _____ **Campsis.**
      3. Leaf scars V-shaped, with 3 bundle scars; terminal bud present; clusters of ovoid capsules to 0.5 cm long; rare species _____ **Decumaria.**
   2. Plants climbing by tendrils or by twisting petioles or by twining.
      4. Plants climbing by twisting petioles; stem and pith with 5 or more angles; clusters of achenes with plumose styles often present_____**Clematis.**
      4. Plants climbing by tendrils or by twining stems; stems rounded to 4-angled; fruits otherwise.
         5. Plants climbing by tendrils attached to persistent petiolar stalks on each side of twig; pith in cross-section with a crosslike pattern; fruit an elongate capsule to 15 cm long by 2.5 cm wide; common species_____**Bignonia.**
         5. Plants climbing by twining, tendrils lacking; pith and fruit otherwise; uncommon to rare species.
            6. Buds scales obvious and long-pointed; bundle scars 3, but usually obscured; bud scales present at twig bases; fruit fleshy (berries)_____ **Lonicera.**
            6. Buds scales obscured or with acute tips; bundle scar 1 and evident; bud scales absent from twig bases; fruit dry (elongate follicles) _____ **Trachelospermum.**
1. Leaf scars alternate.
   7. Tendrils present OR pith brown with woody partitions at the nodes, or both.
      8. Twigs ringed by a line at nodes; tendrils emerging from same side of twig as the bud; tendrils unbranched _____**Brunnichia.**
      8. Twigs lacking a line around nodes; tendrils, if present, on opposite side of twig from bud, and branched (except in *Vitis rotundifolia*).
         9. Tendril tips enlarged into adhesive discs that attach to surfaces; aerial roots sometimes present _____**Parthenocissus.**
         9. Tendril tips pointed and twining, or tendrils absent; aerial roots usually lacking.
            10. Bark shredding into loose strips; pith brown with woody partitions at nodes _____ **Vitis.**

10. Bark tight; pith brown or white, but lacking woody partitions at nodes.
    11. Tendrils unbranched; pith brown or pale, continuous both at nodes and internodes; fruiting inflorescence paniculate (*V. rotundifolia*) _____ **Vitis.**
    11. Tendrils branched; pith white, the older twigs sometimes diaphragmed in internodes; fruiting inflorescence cymose _____ **Ampelopsis.**
7. Tendrils absent and pith otherwise, plants climbing by aerial roots or by twining, or trailing.
    12. Aerial roots present along internodes and often abundant, especially when climbing; twigs brownish; buds evident and naked, red-brown to yellowish; fruit a white drupe in panicles, often present in winter; plants trailing or climbing; ALL PARTS POISONOUS (*T. radicans*) _____ **Toxicodendron.**
    12. Aerial roots absent from internodes (plants may root at nodes or stems tips when in contact with ground); buds and fruits not in above combination (if buds naked, then either buds sunken or twig and bud colors otherwise).
        13. Stems trailing only, wiry and slender (to 3 mm thick), often bristly; buds scaly; petioles incompletely deciduous, the basal portion remaining as a "petiolar stump;" fruits absent in winter (unarmed dewberries) _____ **Rubus.**
        13. Plants chiefly climbing, occasionally trailing, or sprawling, but stems usually much thicker, not wiry; buds various; petioles completely deciduous and "petiolar stumps" absent (elevated leaf scars sometimes present, as in *Akebia* and *Berchemia*); fruit various, sometimes persistent.
            14. Knobby projection present and solitary above leaf scar; lateral buds paired on each side of knob; bundle scars 3; stems and buds hirsute-pubescent _____ **Pueraria.**
            14. Knobby projection absent or not as above; buds, bundle scars, and pubescence not in above combination.
                15. Twigs, 5-angled, hollow (or pith spongy) and inner bark strongly malodorous; bundle scar 1; plants more sprawling than twining, climbing to 3 m high (rarely to 10 m high); fruit a red berry _____ **Solanum.**
                15. Twigs various but lacking above combination of features; bundle scars various; plants obviously twining, high climbing in most (exceptions *Cocculus* and *Menispermum*); fruit not a berry.
                    16. Fruit a legume, often persistent in winter; hornlike projections often present on lower sides of leaf scars; bud two-scaled or silvery pubescent _____ **Wisteria.**
                    16. Fruit otherwise; twigs and buds lacking above combination of features.
                        17. Buds scaly, protruding from twig surfaces, not superposed or sunken.
                            18. Buds ovoid, protruding at right angles from the twig; bundle scar 1; fruit an orange capsule with fleshy seeds, in racemes _____ **Celastrus.**
                            18. Buds acute, either ascending or appressed; bundle scars various; fruit otherwise.
                                19. Leaf scars flat or only slightly elevated from twig surface; terminal buds present; bundle scars 3; fruit a red drupe _____ **Schisandra.**
                                19. Leaf scars on stalklike projections from sides of twigs; terminal buds absent; bundle scars and fruit otherwise.
                                    20. Buds appressed against twig, the scales usually < 5; bundle scar 1 or obscure; fruit a blue drupe _____ **Berchemia.**
                                    20. Buds spreading from twig, the scales > 5; bundle scars 3 or more or obscure; fruit a fleshy pod with many black seeds ____ **Akebia.**
                        17. Buds hairy-naked and/or sunken-superposed.
                            21. Leaf scar horseshoe-shaped, with 3 bundle scars, surrounding a silvery pad bearing several superposed and silvery-hairy buds; fruit a cylindric capsule to 8 cm long _____ **Aristolochia.**
                            21. Leaf scars oval, round, or half-round; buds various; bundle scars many (except in *Cocculus*); fruit a drupe.

22. Leaf scars elliptic, not notched; drupe black, the seed caved in on one side, cup-like; woody vines climbing high in trees ____ **Calycocarpum.**
22. Leaf scars orbicular to kidney shaped, often notched at apex; drupes black or red, the seed crescent-shaped; woody vines climbing to 5 m high.
   23. Twigs lustrous, striate-grooved, often purplish or reddish; leaf scars concave, the buds partly sunken and superposed; drupe black, the seed keeled, shaped like a half-moon _____ **Menispermum.**
   23. Twigs dull green, not grooved; leaf scars flat, the buds emergent and naked; drupe red, the seed with cross-ridges, shaped like a snail _____ **Cocculus.**

# KEY G

**Angiosperms with leaves deciduous; stems erect and unarmed; leaf scars opposite or whorled, and joined around the twig or connected by lines or ridges.** Note: species that have leaf scars or stipular scars almost but not completely touching, such as *Broussonetia papyrifera* and some *Acer*, are included in Key H.

1. Terminal buds absent; all shrubs; pith continuous or porous (*Deutzia*, in which terminal buds may be present or absent, has hollow twigs, and is included under next couplet #1).
   2. Bundle scars clearly > 3 (usually 5 or 6); twigs thick, often > 5 mm in diameter _____ **Sambucus.**
   2. Bundle scars 1 or 3 or uncountable; twigs usually < 5 mm in diameter.
      3. Buds and remnant leaves white-woolly with stellate hairs; fruit a capsule, about 8 mm long, 2-valved, often persistent_____ **Buddleja.**
      3. Buds and remnant leaves (if present) lacking woolly and stellate pubescence; fruit otherwise, or absent.
         4. Bundle scar 1 or obscure; species of wetlands or uplands.
            5. Bark shreddy; twigs winged or angled, often 2-edged; buds obscured or developing into short leafy branches in leaf axils; fruit a capsule, often persistent_____ **Hypericum.**
            5. Bark scaly; twigs rounded; buds various, but rarely developing into leafy branches; fruit otherwise.
               6. Shrub to 3 m tall; buds barely emergent; pith continuous; leaf scars opposite or whorled; fruit a ball-like cluster of nutlets, sometimes persistent; lowland species, often in standing water _____ **Cephalanthus.**
               6. Shrub to 2 m tall; buds evident and scaly; pith porous or continuous; leaf scars opposite only; fruit fleshy, a red or white berry, clustered in leaf axils or on short stalks and persistent in winter; upland species_____ **Symphoricarpos.**
         4. Bundle scars clearly 3; species of uplands and disturbed areas.
            7. Buds with 8 or more dark-tipped scales; dried stipules often present; fruits a cluster of black drupes surrounded by jagged sepals; non-native species escaping from cultivation _____ **Rhodotypos.**
            7. Buds otherwise; stipules absent; native species, usually rare and local.
               8. Lateral buds visible and spreading, or hidden behind leaf scar, with green buds breaking through leaf scar in late winter; capsules ovoid, about as long as wide; shrubs to 3 m tall _____ **Philadelphus.**
               8. Lateral buds visible but appressed to twig and loosely scaled; capsules cylindric, longer than wide; shrubs usually < 2 m tall _____ **Diervilla.**
1. Terminal buds present; shrubs and trees.
   9. Pith hollow or excavated; shrubs.
      10. Bark very loose, exfoliating in mature stems; twigs with stellate pubescence; fruit a capsule, often persistent, in terminal clusters_____ **Deutzia.**
      10. Bark furrowed to splitting, but not typically exfoliating; twigs glabrous or pubescent with simple hairs; fruit a fleshy berry in leaf axils_____ **Lonicera.**

9. Pith continuous; shrubs or trees.
    11. Bundle scars > 5 and not in 3 groups.
        12. Shrubs; terminal buds about 2 cm long, hairy, the outer bud scales elongate and as long as the buds (*H. quercifolia*) _____**Hydrangea.**
        12. Trees; terminal buds to 1.5 cm long, smooth or hairy, but lacking elongate outer scales.
            13. Terminal bud acute, to 1.5 cm long, with many imbricate scales; leaf scars shield-shaped _____**Aesculus.**
            13. Terminal buds blunt, < 1 cm long; the bud scales in pairs; leaf scars half-round to U-shaped _____**Fraxinus.**
    11. Bundle scars 3–5, or more and in 3 groups, or obscure.
        14. Buds naked and stalked, pubescent and veiny, to 3 cm long; restricted to higher elevations of KY and TN (*V. lantanoides*)_____**Viburnum.**
        14. Buds scaly (pubescent in some or appearing naked but not veiny), mostly < 2 cm long (some flower buds may be longer); plant distributions various.
            15. Biscuit-shaped flower buds present, these 4-scaled; vegetative buds 2-scaled and pointed; small trees (*C. florida*)_____**Cornus.**
            15. Biscuit-shaped flower buds absent; vegetative buds various; shrubs or trees.
                16. Leaf scars angling upward and meeting at a point higher than tips of adjacent lateral buds, forming a toothlike projection; buds blunt and gray-woolly; twigs green; bark tightly furrowed; samaras often present in winter (*A. negundo*)_____**Acer.**
                16. Leaf scars meeting at a point lower than tips of adjacent lateral buds, or if so, then not forming a toothlike projection; buds, twigs, and bark not in above combination; fruit various, with samaras being produced in other *Acer*, but these usually not persistent.
                    17. Vegetative buds with 2 evident scales or appearing naked.
                        18. Twigs dull-colored, usually gray or brown; terminal buds scurfy (with powdered or rough-scaly appearance, usually reddish or brownish) or waxy with purplish or grayish bloom, usually long and slender, and the larger flower buds swollen at base _____**Viburnum.**
                        18. Twigs brightly colored, usually green or red; terminal buds smooth to pubescent, but not scurfy or waxy, conical to ovoid, and not swollen at base of flower buds.
                            19. Twigs swollen at nodes (due to raised leaf scars); vegetative buds mostly < 5 mm long, conical, flattened and unstalked, often appressed against the twig; shrubs; fruit a drupe _____**Cornus.**
                            19. Twigs not swollen at nodes; vegetative buds mostly > 5 mm long, ovoid, round-tipped and often stalked, small trees; fruit a samara_____**Acer.**
                    17. Vegetative buds with 4 or more visible scales.
                        20. Shrubs, mostly < 3 m tall, bark shredding into loose strips in mature individuals; twigs often with vertical ridges; bud scales often persistent at bases of young twigs; leaf scars often raised (nodes swollen), but less so in *Hydrangea*.
                            21. Shrubs to 2 m tall; twigs with strong ridge running vertically down from middle of leaf scar; pith < ½ diameter of twig; bundle traces obscured; fruit fleshy, a pair of berries, stalked from the leaf axil, but absent in winter; species rare in E TN (*L. canadensis*) _____**Lonicera.**
                            21. Shrubs to 3 m tall; twigs lacking strong vertical ridge from leaf scar, but with several vertical ridges evident; pith often relatively large, > ½ diameter of twig; bundle traces evident; fruit a capsule; species more widespread in KY and TN.
                                22. Leaf scars more or less flush with twig surface; bundle scars located near edge of leaf scar; capsules to 2 mm long, broader than long, numerous in open flat-topped clusters _____**Hydrangea.**

22. Leaf scars raised, strongly projecting from sides of twigs; bundle scars located within edges of leaf scar; fruit dry, a capsule, 5–9 mm long and broad, usually fewer than 6 in a cluster, often persistent _____**Philadelphus.**
20. Shrubs or trees; bark otherwise, not shredding; bud scales deciduous, not persistent at bases of twigs.
23. Leaf scars shield-shaped or triangular; twigs usually > 5 mm thick; buds ovoid, often > 1 cm long, sharp-pointed; fruit a leathery capsule with large black seeds, small to large trees _____**Aesculus.**
23. Leaf scars U- or V-shaped; twigs usually < 5 mm thick; buds much smaller, usually lanceolate or globose, blunt or sharp pointed; fruit a drupe or samara; shrubs or trees.
24. Shrubs; twigs gray or brown, often pubescent or roughened; bud scales acute, usually fewer than 8; lateral buds usually appressed against the twig; fruit a drupe _____**Viburnum.**
24. Trees; twigs red or brown or green, smooth or pubescent; bud scales various, if acute then 8 or more; lateral buds mostly spreading from twig; fruit a samara_____**Acer.**

# KEY H

**Angiosperms with leaves deciduous; stems erect and unarmed; leaf scars opposite or whorled, and not joined around the twig or connected by lines or ridges.** Note: including species with stipular scars nearly but not completely touching.

1. True end buds lacking, the false end buds often paired at twig apex, or not evident (Note: species of *Aesculus* have true end buds, but some twigs may have paired end buds due to the presence of a fruit stalk scar at the twig tip; these species, with buds multi-scaled and over 1 cm long, unlike any species in this category, are keyed under the next couplet).
2. Bundle scar 1 or 3; leaf scars various, somewhat circular only in *Decodon*; shrubs.
3. Bundle scar 3; fruits dry and bladdery, these often persistent.
4. Twigs green, lacking odor; stipular scars present; buds scaly _____**Staphylea.**
4. Twigs brown, with spicy odor; stipular scars absent; buds dark and lacking evident scales _____**Calycanthus.**
3. Bundle scar 1; fruit otherwise.
5. Shrubs with arching stems, rising to 3 m tall, the twigs green, angular, and rooting at the tips; buds sunken and not readily evident; wetland plants usually in standing water _____**Decodon.**
5. Shrubs with erect branches, not arching and rooting; buds evident; twigs gray-green to brown; upland or lowland plants, but not in standing water.
6. Bark rough; buds dark purple, ovoid, about 7 mm long, the scales strongly keeled; non-native shrubs, occasionally escaping; capsules persisting into winter_____**Syringa.**
6. Bark smooth; buds otherwise, pale to dark brown, rare native shrubs parasitic on tree roots; fruit a greenish drupe, but rarely present in winter.
7. Buds > 5 mm long, long-pointed, pale brown; plants to 3 m tall _____**Buckleya.**
7. Buds to 3 mm long, ovoid to acute, dark brown; plants < 1.5 m tall _____**Nestronia.**
2. Bundle scars > 3; leaf scars circular or nearly so; trees.
8. Stipular scars present and almost touching; twigs bristly-hairy; bundle scars about 5; milky sap may be present_____**Broussonettia.**
8. Stipular scars absent; twigs smooth; bundle scars many in an ellipse; milky sap absent.

9. Leaf scars whorled; pith continuous; capsules elongate, cigar-shaped; flower buds absent _____ **Catalpa.**
9. Leaf scars opposite; pith chambered or excavated; capsules ovoid; flower buds present, brown-pubescent, in large panicle _____ **Paulownia.**
1. True end buds present.
    10. Buds naked and stalked, sometimes superposed; twigs stellate pubescent, slightly 4-angled, with rank odor; leaf scars crescent-shaped, the single bundle scar U-shaped, fruits drupelike, purplish, often persistent in axillary clusters; shrubs _____ **Callicarpa.**
    10. Buds scaly; twigs lacking stellate pubescence; other features not in above combination; shrubs or trees.
        11. Bundle scars clearly > 1; small to large trees.
            12. Bundle scars 3 or in 3 groups; leaf scars U- or V-shaped or shield-shaped; bud scales many; twigs rounded.
                13. Leaf scars U- or V-shaped; buds < 1 cm long _____ **Acer.**
                13. Leaf scars shield-shaped; buds often > 1 cm long _____ **Aesculus.**
            12. Bundle scars many in a curved line; leaf scars half-round to U-shaped; bud scales in 2 or 3 pairs; twigs rounded or angled_____ **Fraxinus.**
        11. Bundle scar 1, dot-like, curved, U-shaped or obscure; shrubs or small trees.
            14. Buds solitary or multiple with several collateral buds, and larger flower buds often present; twigs with conspicuous warty lenticels; pith diaphragmed or chambered or excavated; twigs often 4-angled; yellow flowers appearing in late winter_____ **Forsythia.**
            14. Buds solitary at nodes, or sometimes with superposed buds; lenticels inconspicuous; pith continuous or spongy in *Euonymus*; twigs various, sometimes angled; flowering in spring and summer.
                15. Twigs green and waxy, often 4-angled or with corky wings; fruit a capsule, rounded or lobed, the seeds with an orange or red fleshy covering (aril)_____ **Euonymus.**
                15. Twigs brown or gray, not waxy or winged, but sometimes angled; fruit a blackish or bluish drupe.
                    16. Twigs angled; leaf scars at least 2 mm wide; bud scales keeled and awned; terminal bud > 3 mm long _____ **Chionanthus.**
                    16. Twigs rounded; leaf scars < 2 mm wide; bud scales various, the scales sometimes keeled or pointed; terminal bud < 3 mm long.
                        17. Lateral buds single in leaf axils; terminal bud ovoid; common plants, in cultivation or naturalized, often with remnant leaves_____**Ligustrum.**
                        17. Lateral buds several above leaf axil (superposed); terminal buds acute; rare native plants of swamps or cedar glades, rarely with remnant leaves _____ **Forestiera.**

# KEY I

Angiosperms with leaves deciduous; stems erect and unarmed; leaf scars alternate; catkins or catkinlike inflorescences (densely flowered racemes or spikes) present in winter.

1. Shrubs to 2 m tall, stems with pungent aroma when crushed; lateral buds very small (usually < 3 mm) or hidden.
    2. Lateral buds hidden by raised leaf scars; bundle scars many; fruit a red drupe; upland shrubs of limestone areas (*R. aromatica*)_____ **Rhus.**
    2. Lateral buds visible above leaf scar; bundle scars 3; fruit a bristly bur; lowland shrubs, restricted to stream margins and gravel bars near KY–TN border _____**Comptonia.**
1. Trees or shrubs, usually much taller; pungent odor lacking (but wintergreen odor in some *Betula*); buds larger and evident.
    3. Bundle scar 1; inflorescence not a true catkin, actually a reddish raceme of bisexual, bracteate flower buds loosely arranged on the axis; shrubs to 4 m tall _____ **Leucothoe.**

3. Bundle scars 3 or obscure; inflorescence of true catkins, unisexual and bracteate, with the flowers densely packed on the axis; shrubs or trees.
    4. Pistillate catkins woody and conelike; buds stalked (with a necklike constriction at base) in the widespread species _____**Alnus.**
    4. Pistillate catkins soft-scaly, not conelike; buds not stalked.
        5. Trees; spur branches (short stout twigs with numerous crowded leaf scars, often topped by a bud) present; bark often banded horizontally with elongate lenticels, becoming scaly and peeling in some species; twigs with wintergreen aroma in some species _____ **Betula.**
        5. Shrubs or small trees; spur branches absent; bark otherwise; twigs lacking wintergreen odor.
            6. Small trees; bark rough, shredding into loose strips; buds lanceolate and acute, the bud scales > 6, with fine striations; twigs lacking red glandular hairs _____**Ostrya.**
            6. Shrubs; bark otherwise; buds ovoid and obtuse, the bud scales ≤ 6, and lacking striations; twigs in the more widespread species (*C. americana*) with reddish glandular hairs _____ **Corylus.**

# KEY J

**Angiosperms with leaves deciduous; stems erect and unarmed; leaf scars alternate; twigs with lines > ¾ encircling the stem at each node (from encircling stipules or nodal sheaths), OR leaf scars > ½ encircling the twig and inner bark yellow.**

1. Shrub < 1 m tall.
    2. Stems jointed, encircled by nodal sheaths; inner bark not yellow; fruit a triangular nutlet ___ **Polygonella.**
    2. Stems not jointed, more than ½ encircled by narrow leaf scars; inner bark yellow; fruit of follicles _____ **Xanthorhiza.**
1. Trees.
    3. True end bud absent.
        4. Leaf scars encircling buds; stipular scars completely encircling twigs; large trees with whitish inner bark; twigs smooth _____**Platanus.**
        4. Leaf scars rounded, located below the buds; stipular scars only partly encircling the twigs; small trees with shallowly ridged bark; twigs hairy _____**Broussonetia.**
    3. True end bud present.
        5. Stipular scars encircling about ¾ the circumference of the twig; terminal buds narrow and slender, multi-scaled; pith continuous; bark smooth and gray; fruit a triangular nut in a husk_____**Fagus.**
        5. Stipular scars completely encircling the twig; terminal buds ovoid or blunt, bud scales 1 or 2; pith diaphragmed; bark varying from smooth to furrowed; fruit a conelike aggregate, often persistent.
            6. Bud scale 1, pubescent or glabrous, ovoid; aggregates of follicles, these remaining on the elongate receptacle _____**Magnolia.**
            6. Bud scales 2, glabrous, flattened and similar to a duckbill; aggregates of woody samaras, these falling from the elongate receptacle _____**Liriodendron.**

# KEY K

**Angiosperms with leaves deciduous; stems erect and unarmed; leaf scars alternate; pith chambered or diaphragmed or spongy-porous.** Note: including only those species that have chambered or diaphragmed pith through the internodes; for example, paper mulberry (*Broussonetia papyrifera*) and others may have a thin diaphragm at the nodes only, and are excluded here.

1. Shrubs, many-branched, 2–3 m tall; terminal buds present.
    2. Pith chambered; bark smooth; terminal buds naked or valvate-scaly; leaf scars half-circular or kidney shaped; fruit dry, a 2-valved, hairy capsule, to 7 mm long, borne in long racemes, these persistent into winter _____**Itea.**

2. Pith spongy with age; bark often shredding into loose strips; terminal buds imbricate-scaly; leaf scars V-shaped, narrow and often half-encircling the twig; fruit fleshy, a berry, but usually absent in winter _____ **Ribes.**
1. Small to large trees, much taller at maturity, usually with single main stem; terminal buds various.
   3. Terminal buds false.
      4. Leaf scars in > 1 plane; bundle scar 1, crescent-shaped; lateral buds divergent from twig; bark blocky in mature plants _____ **Diospyros.**
      4. Leaf scars in one plane; bundle scars 3 or obscure; lateral buds appressed against twig; bark smooth or warty in mature specimens _____ **Celtis.**
   3. Terminal buds true.
      5. Bundle scar 1; species rare or local in Appalachian regions.
         6. Buds silvery pubescent; pith spongy-porous _____ **Stewartia.**
         6. Buds glabrous or with brownish or grayish pubescence; pith chambered or diaphragmed.
            7. Small trees to 10 m tall; bark grayish with brownish or greenish ridges, sometimes warty or with vertical stripes; twigs sometimes hairy; rounded flower buds often present in winter _____ **Symplocos.**
            7. Trees to 30 m tall; bark purplish-scaly on mature specimens, white-striped when young; twigs smooth; flower buds absent in winter _____ **Halesia.**
      5. Bundle scars 3 or more; species common in the region (*Pyrularia* restricted to Appalachians).
         8. Bundle scars 3 (many bundle scars in branch scars of *Pyrularia*); buds with hard shiny scales, green and purple to dark brown in winter; superposed buds absent.
            9. Pith diaphragmed; circular branch scars absent; buds to 8 mm long; large tree _____ **Nyssa.**
            9. Pith spongy; circular branch scars present; buds to 12 mm long; shrub _____ **Pyrularia.**
         8. Bundle scars > 3; buds lacking hard shiny scales, gray to reddish-brown; buds often superposed.
            10. Terminal buds gray, rounded; pith chambered; twigs with odor of walnut; leaf scars in > 1 plane; large tree _____ **Juglans.**
            10. Terminal buds red-brown, slender and pointed; pith diaphragmed, especially in older twigs; twigs with peppery or skunklike odor; leaf scars in one plane; small tree _____ **Asimina.**

# KEY L

Angiosperms with leaves deciduous; stems erect and unarmed; leaf scars alternate; pith continuous; bundle scar 1.

1. Shrubs with branches bright green, strongly angled and flexible _____ **Cytisus.**
1. Shrubs or trees lacking branches as described above.
   2. Bud naked or valvate-scaly or with scales obscured by dense pubescence.
      3. Twigs with dark-centered scales and/or stellate pubescence.
         4. Twigs and buds scaly, the scales with dark centers and silvery or brownish edges; stellate hairs sometimes present also, especially on the leaves; fruit fleshy, drupelike, red or yellow, often silvery-scaly; invasive exotic species _____ **Elaeagnus.**
         4. Twigs and buds lacking silvery scales, but stellate hairs present; fruit otherwise; native species, infrequent to rare.
            5. Terminal buds absent, the buds brownish and blunt, densely hairy, often superposed; fruit a dry drupe to 10 mm long _____ **Styrax.**
            5. True end buds present, the buds acute and silvery, usually not superposed; fruit a small capsule, about 5 mm long, in racemes _____ **Clethra.**
      3. Twigs lacking dark-centered scales and stellate pubescence.
         6. Buds naked or hairy; low shrubs often < 1 m tall.
            7. Stem slender, few-branched, wandlike, often white-hairy, bearing terminal clusters of small follicles; species mostly of moist open sites (*S. tomentosa*) _____ **Spiraea.**

7. Stems forming a low rounded crown, much-branched, green and smooth, bearing disclike capsules, persistent into winter; species of uplands and stream banks _____ **Ceanothus.**
6. Buds valvate-scaled, often reddish and appressed against the twig; shrubs > 1 m tall.
   8. Twigs yellowish; fruit a dry capsule in racemes, usually persistent _____ **Lyonia.**
   8. Twigs greenish, brownish, or reddish; fruit a berry, absent in winter.
      9. Twigs glabrous or hairy in lines; buds all 2-scaled (*V. erythrocarpum*) _____ **Vaccinium.**
      9. Twigs generally pubescent; flower buds 4-scaled and vegetative buds 2-scaled (*G. ursina*) _____ **Gaylussacia.**
2. Bud scales evident and imbricate and not obscured by dense pubescence.
   10. True end buds absent.
      11. Trees; buds dome-shaped and obtuse; fruit a capsule to 7 mm long in drooping clusters _____ **Oxydendrum.**
      11. Shrubs; buds and fruit not in above combination.
         12. Stem often few-branched, and bearing clusters of small follicles, these persistent into winter; mostly uncommon species of open, wet sites _____ **Spiraea.**
         12. Stems multi-branched, bearing fleshy fruits, but these usually not persistent into winter.
            13. Twigs speckled with tiny wartlike structures, often greenish, smooth or hairy (*V. corymbosum* and relatives) _____ **Vaccinium.**
            13. Twigs lacking speckled surface, the color various, often brown or reddish.
               14. Buds scales (and remnant leaves) with yellow resin dots; low shrubs mostly < 2 m tall _____ **Gaylussacia.**
               14. Bud scales and leaves lacking yellow resin dots; shrubs often > 2 m tall _____ **Vaccinium.**
   10. True end buds present.
      15. Twigs green, with spicy-aromatic odor _____ **Sassafras.**
      15. Twigs lacking above combination of features.
         16. Buds relatively small, to 5 mm long; spur branches often present on older twigs; bud scales 4–6, smooth or hairy, rounded with smooth or erose margins; tiny stipular scars present; fruit fleshy, berrylike, but absent in winter.
            17. Bud scales pointed; buds often superposed; stalk of fruit < 8 mm long (except in *I. longipes*) _____ **Ilex.**
            17. Bud scales blunt; buds rarely superposed; stalk of fruit > 8 mm long ___ **Nemopanthus.**
         16. Buds relatively large, often > 5 mm long; spur branches absent; bud scales > 6; with white-ciliate margins; stipular scars absent; fruit a dry capsule, often persistent.
            18. Bark smooth to scaly, but not exfoliating; bud scales often keeled and with mucronate tips; capsule cylindric and 5-locular, > 1 cm long, the peduncle about equaling or shorter than the fruit; plants of medium to high elevations of the Appalachians _____ **Rhododendron.**
            18. Bark on older stems exfoliating; bud scales rounded on back and pointed or rounded at apex; capsule ovoid and 4-locular, < 1 cm long, bristly, the peduncles many times longer than the fruit; high-elevation species of E TN _____ **Menziesia.**

# KEY M

**Angiosperms with leaves deciduous; stems erect and unarmed; leaf scars alternate; pith continuous; bundle scars 3 or more or obscure; true end buds present or end buds clustered (as in *Quercus* and *Hibiscus*).**

1. Buds clustered at twig tips (and with clustered leaf scars in *Hibiscus*), with several similar-sized end buds evident (end buds may be obscured by other plant parts in *Hibiscus*); bundle scars > 3 (may be obscure in *Hibiscus*); fruit an acorn, or a large capsule to 3 cm long.

2. Shrubs; terminal buds obscured by remains of branches and twig tips and persistent stipules; pith round in cross-section; fruit a capsule about 3 cm long, usually persistent _____ **Hibiscus.**
2. Trees; terminal buds evident; pith star-shaped in cross-section; fruit an acorn _____ **Quercus.**
1. Buds usually solitary at twig tips (if clustered then bundle scars clearly 3); fruit various, but not an acorn or a large capsule to 3 cm long.
    3. Bud scales long-pointed and black-tipped; bark with sticky sap and strong pinelike or menthol odor; leaf scars half-round, with bundle scars 3 or in groups of 3; small tree of restricted distribution in S TN _____ **Cotinus.**
    3. Buds and bark and leaf scars not in above combination; mostly common species (except for *Sorbus*, *Fothergilla* and some *Ribes* species).
        4. Bundle scars > 3 (may be in 3 groups) or if bundle scars 3, as in some *Fagus*, then buds 15–25 mm long, very sharp-pointed, and over 10-scaled.
            5. Bud scales imbricate; buds relatively large or elongate, often 1 cm or more long.
                6. Small trees to 12 m tall; bundle scars 5; leaf scars narrowly linear; terminal buds scales purplish and resinous; fruit a red pome; high-elevation species of E TN _____ **Sorbus.**
                6. Tall trees; bundle scars various; bud scales otherwise; leaf scars elliptic to shield-shaped or 3-lobed; more widespread species.
                    7. Buds ovoid; leaf scars shield-shaped to 3-lobed; fruit a hickory nut _____ **Carya.**
                    7. Buds narrowly lanceolate, very sharp-pointed; leaf scars elliptic; fruit a beech nut ___ **Fagus.**
            5. Bud scales valvate or lacking (buds naked); buds and twigs usually smaller than described above.
                8. Small trees; buds naked, dark brown and hairy, oblong; globose flower buds often present; leaf scars in 1 plane; bark smooth and gray _____ **Asimina.**
                8. Shrubs to tall trees; buds otherwise, if naked then yellow-brown; globose flower buds absent; leaf scars in > 1 plane; bark various.
                    9. Shrubs or small trees; buds naked and yellow-brown or reddish and valvate-scaled; lateral buds solitary; bark smooth in tree species; fruit a white drupe in panicles; ALL PARTS POISONOUS! _____ **Toxicodendron.**
                    9. Large trees; buds yellow, gray, brown, or red-brown, valvate-scaled; lateral buds often superposed; fruit a nut with dehiscent husk; bark scaly in mature specimens _____ **Carya.**
        4. Bundle scars 3 and buds not as described above for *Fagus*.
            10. Buds naked, or appearing naked or valvate-scaly after the outer bud scales drop off; twigs and bark lacking distinct odor.
                11. Bud unstalked and naked, to 5 mm long; twigs lacking stellate trichomes; leaf scars in > 1 plane; fruit berrylike _____ **Frangula.**
                11. Bud stalked (constricted at the base), appearing naked or with a few outer scales (soon deciduous), 5–10 mm long; twigs with stellate pubescence; leaf scars in 1 plane; fruit capsular.
                    12. Shrubs often > 3 m tall; buds flattened, lacking remnant outer scales; flowering in the fall and winter; widespread in region _____ **Hamamelis.**
                    12. Low shrubs to 2.5 m tall; buds more rounded or slightly compressed, often with broken remnants of larger outer scales; flowering in the spring; species restricted to E TN _____ **Fothergilla.**
            10. Terminal buds imbricate scaly and unstalked (flower buds stalked in some species); twigs and bark with notable odor in some genera.
                13. Bark shredding into loose strips; twigs lined or ridged; all shrubs.
                    14. Inner bark lacking odor; buds lacking resin dots; upper two bundle scars smaller than the lower one; pith tan; fruits of inflated follicles, these often persistent into winter; plants to 3 m tall _____ **Physocarpus.**
                    14. Inner bark aromatic; bud scales with resin dots; pith white; bundle scars all alike; fruit a black berry; plants < 2 m tall (*R. americanum*) _____ **Ribes.**

13. Bark otherwise; twigs various; shrubs or trees.
    15. Shrubs to 1 m tall; scraped twig with skunklike odor; high-elevation species of E TN (*R. glandulosum*) _____ **Ribes.**
    15. Taller or sprawling shrubs (< 1 m high in *Prunus pumila*), or trees; scraped twig lacking skunklike odor; plants of various habitats.
        16. Twigs and bark with spicy-aromatic odor (not bitter almond as in *Prunus*); other distinctive features as described below.
            17. Trees with ridged and furrowed bark; twigs greenish (or red-brown); end buds ovoid, 5–10 mm long, bundle scar a curved line or broken into 3 sections; stalked flower buds absent _____ **Sassafras.**
            17. Shrubs with bark smooth or dotted with lenticels; twigs greenish-brown; buds lanceolate, < 5 mm long, often superposed; bundle scars of 3 dots; flower buds present, these globose and stalked and clustered at nodes _____ **Lindera.**
        16. Twigs and bark otherwise; plants lacking other combination of features described above.
            18. Shrubs or small trees (> 5 m tall only in *Amelanchier*); leaf scars narrow, V- or U-shaped, about as wide as the twig; lateral buds often nearly as large as the terminal buds, appressed or curved toward the sides of the twigs, sometimes twisted; fruit a berry or pome.
                19. Shrubs with many arching stems and branches forming an irregular crown; fruit, if present, a berry; very rare plants of C TN (*R. aureum*) _____ **Ribes.**
                19. Shrubs with erect or trailing stems, or small trees, fruit a pome; rare to generally common species of wide distribution.
                    20. Terminal buds rounded in cross-section, with 6 visible scales, the bud tip appearing twisted; second lowest bud scale less than half as long as the bud; shrubs trailing or erect _____ **Amelanchier**
                    20. Terminal bud flattened, with 4 or 5 visible scales; second lowest bud scale over half as long as the bud; shrubs erect _____ **Aronia.**
            18. Small to large trees (most species reaching heights of > 7 m tall, with exception of *Cornus alternifolia* and some *Prunus* spp.); leaf scars otherwise, usually half-round, elliptic, lobed, or triangular; fruit various.
                21. Buds with lowest bud scale directly above leaf scar and this basal scale as wide as the twig; pith angled in cross-section _____ **Populus.**
                21. Bud scales otherwise; pith usually rounded in cross-section.
                    22. Buds glossy, green or purple-brown or green with dark marginal bands, lanceolate to ovate, 5–12 mm long; twigs often with corky wings; bundle scars white-margined; fruit spherical, a ball of beaked capsules _____ **Liquidambar.**
                    22. Buds, twigs, and fruit otherwise, not in above combination; twigs lacking corky wings.
                        23. Terminal bud smooth, the visible scales 2–4 and smooth; lateral buds very small or obscured by leaf scars; leaf scars raised and crowded toward stem tip; bark with unpleasant odor; the terminal twig sometimes exceeded in length by a lower lateral branch (*C. alternifolia*) _____ **Cornus.**
                        23. Terminal buds otherwise, either bud scales > 4 or bud scales hairy (or both); lateral buds usually evident; leaf scars various, but if crowded toward apex, then lateral buds evident and clustered; odor present only in *Prunus* (bitter almond); terminal twig rarely exceeded by a lateral branch.

24. Twigs with odor of bitter almonds; bark in older trees platy, gray or blackish, in saplings the bark smooth with horizontal stripes; buds ovoid and acute, often collateral or clustered at apex; fruits, if present, 1-seeded; small to large trees; (cherries and peach) _____**Prunus.**

24. Twigs lacking almond odor; bark and other features not in above combination; fruits, if present, many seeded; small trees.
    25. Buds blunt, gray and pubescent, to 7 mm long; trees with rounded crowns, the major branches spreading ___ **Malus.**
    25. Buds acute, glabrous, or if pubescent, then > 8 mm long; trees often with upswept branches, producing a narrow crown _____**Pyrus.**

# KEY N

Angiosperms with leaves deciduous; stems erect and unarmed; leaf scars alternate; pith continuous; bundle scars 3 or more or obscure; true end buds absent.

1. Buds not clearly evident, either buried under leaf scar or sunken nearly level with twig surface, and frequently superposed; bundle scars 3–5 or obscure; fruit a multi-seeded legume, either 10–40 cm long and purplish-brown OR to 10 cm long and splitting into two halves with dark seeds attached.
    2. Bark tightly furrowed in mature specimens; buds all buried in leaf scar; twigs slender, with or without stiff bristles; legumes grayish or brownish, to 10 cm long, lacking pulp, splitting apart while on the tree, with dark seeds, these to 5 mm long _____**Robinia.**
    2. Bark with loose scales or plates; buds sunken, but at least some buds located above leaf scar; twigs varying from slender to very thick (5–10 mm thick); twigs lacking stiff bristles; legume > 10 cm long, purplish or dark brown, pulpy inside, usually remaining intact while on the tree, the seeds > 5 mm long.
        3. Twigs to 10 mm thick, lacking swollen nodes; buds sunken in hairy pit, about level with twig surface; legume woody (thick-walled), 10–20 cm long, not twisted; seeds 1–2 cm long _____**Gymnocladus.**
        3. Twigs thinner, the nodes usually swollen; buds tiny and knoblike, but slightly projecting from twig surface; legumes not woody (thin-walled), to about 40 cm long, often twisted; seeds about 1 cm long (unarmed forms of honeylocust) _____**Gleditsia.**
1. Buds evident (may be partly sunken in *Amorpha*, or sometimes small and domelike or very hairy), fruit various, but if a legume, then unlike those described above.
    4. Buds naked, usually with dense pubescence, lacking hard scales.
        5. Leaf scars horseshoe-shaped, encircling the buds, or nearly so.
            6. Twigs often about 10 mm thick, and milky sap usually present; remnant panicles of red drupes often present _____**Rhus.**
            6. Twigs usually < 5 mm thick, lacking milky sap; fruit otherwise or absent.
                7. Buds silvery-hairy; plant either with leathery bark or citrus odor; fruit a drupe or a samara.
                    8. Twigs with leathery bark, very flexible and almost unbreakable, lacking citrus odor; bundle scars 5; fruit a greenish drupe about 1 cm long _____**Dirca.**
                    8. Twigs lacking leathery bark, with citrus odor; bundle scars 3; fruit a rounded samara _____**Ptelea.**
                7. Buds brown-hairy, plants lacking leathery bark and citrus odor; fruit a legume often with rim of stamens at base _____**Cladrastis.**
        5. Leaf scars half-circular to 3-lobed, positioned below the buds and < ½ encircling the buds.
            9. Twigs glabrous; drupes white and poisonous; leaf scars 3-lobed, with 3 U-shaped bundle scars; trees _____**Melia.**

9. Twigs pubescent; drupes red and nonpoisonous; terminal clusters of red drupes often persistent; leaf scars crescent to half-round, often with > 3 bundle scars; usually shrublike (*R. copallinum*) ............................................................................................................. **Rhus.**
4. Buds with hard scales, either 1-scaled, valvate-scaly, or imbricate-scaly, the scales smooth or variously pubescent.
   10. Buds with 1 visible scale ............................................................................................................. **Salix.**
   10. Buds with > 1 visible scale.
      11. Twigs and bark with spicy-aromatic odor (not wintergreen as in some *Betula* or bitter almond as in *Prunus*); vegetative buds ovoid and appressed to twig, often superposed; flower buds in clusters at nodes, the buds globose and stalked ............................................... **Lindera.**
      11. Twigs and bark lacking spicy odor; vegetative and flower buds lacking above combination of features.
         12. Shrubs < 4 m tall, the branches very slender and elongate, the plants with "bushlike" or "blackberry-like" habit, in some cases with 1-seeded or 2-seeded, indehiscent legumes.
            13. Stems with hard, persistent "petiolar stumps" remaining on the twigs (stalked glandular hairs and exfoliating bark in *R. odoratus*); unarmed blackberries and raspberries ............................................................................................................. **Rubus.**
            13. Stems lacking hard "petiolar stumps" (flexible bases of petioles persistent in *Neviusia*); stalked glandular hairs absent; bark various (slightly exfoliating only in older stems of *Neviusia*).
               14. Fruit a legume, indehiscent, with 1 or 2 seeds, < 12 mm long, usually persistent in winter; lateral buds often superposed or collateral; stubby lateral projections from twigs absent.
                  15. Legume densely pubescent, strongly flattened, 1-seeded; shrubs to 3 m tall with many slender, barely woody branches, the upper parts often dying back in winter; stems lacking odor of kerosene; stipules usually deciduous _____ **Lespedeza.**
                  15. Legume gland-dotted or sparsely pubescent, slightly swollen or curved, with 1 or 2 seeds; stout shrubs to 4 m tall, stems sometimes with odor of kerosene; stipules often persistent and peglike ............................................... **Amorpha.**
               14. Fruit otherwise, usually absent in winter; lateral buds various, sometimes collateral but rarely superposed; twigs often with stubby lateral projections (spur branches in *Kerria* and broken petiole bases in *Neviusia*).
                  16. Stems brownish; flowers in early spring with conspicuous white stamens; very rare native species of south-central TN ............................................... **Neviusia.**
                  16. Stems bright green; flowers in early spring with yellow showy petals; non-native species, rarely escaping from cultivation ............................................... **Kerria.**
         12. Small to large trees, > 4 m tall (except in some *Castanea*, *Celtis*, *Prunus*, and *Rhamnus*, these with branches thick and short, and lacking a "bushlike" or "blackberry-like" habit), in no cases with 1- or 2-seeded, indehiscent legumes.
            17. Twigs with superposed buds; scraped bark with odor of green beans; small to medium trees to 10 m tall; fruit a legume with > 2 seeds and dehiscent, often persistent.
               18. Twigs grayish or greenish; leaf scars 3-lobed, in > 1 plane; stipules often present, peglike; legumes yellowish, to about 15 cm long ............................................... **Albizia.**
               18. Twigs reddish to purplish, notably white-speckled; leaf scars triangular, with a fringe of red hairs on upper edge, the leaf scars raised and with 2 lines extending down from base, in one plane; stipular scars absent; fruits purplish, to 10 cm long ............................................................................................................. **Cercis.**
            17. Twigs lacking superposed buds; scraped bark lacking odor of green beans (other odors described below); fruit otherwise.

19. Stipular scars absent; twigs either very stout (> 10 mm thick) OR with lenticels elevated, orange and ringed by pale ridges; buds with 2 or 4 visible scales, appearing valvate; leaf scars more or less shield-shaped.
   20. Buds pointed, with 2 scales usually visible; lenticels orange with pale rings; twigs < 10 mm thick; scraped bark lacking a distinct odor; leaf scars shield-shaped, with bundle scars clustered near the center; fruit an inflated capsule with black seeds _____ **Koelreuteria.**
   20. Buds rounded or depressed, with about 4 scales visible; lenticels of pale openings; twigs stout, often > 10 mm thick, the scraped bark with unpleasant, "burned peanut butter" odor; leaf scars shield-shaped, with about 9 bundle scars; fruit a samara _____ **Ailanthus.**
19. Stipular scars present; twigs more slender and lacking orange-ringed lenticels; bud scales overlapping, not appearing valvate (if < 5-scaled, then leaf scar shapes otherwise).
   21. Twigs and trunk with lenticels elongated horizontally, producing a striped appearance of the bark, the bark smooth to scaly or flaky; twigs usually with spur branches; species notable either for distinct odor (bitter almond in *Prunus* or wintergreen in some *Betula*) or for bark peeling away in loose, paper-like sheets (some *Betula*).
      22. Leaf scars in 1 plane; twigs with wintergreen odor or odor lacking; buds 2–8 scaled; trees _____ **Betula.**
      22. Leaf scars in > 1 plane; twigs with bitter-almond odor; buds various, but often multi-scaled; shrubs or small trees _____ **Prunus.**
   21. Twigs and trunk lacking elongated horizontal lenticels; bark various, often ridged and furrowed, but smooth in *Carpinus*, warty in *Celtis*, and flaky in *Planera*; species lacking wintergreen or bitter-almond odor or paper-like bark; twigs lacking spur branches (except in *Maclura*).
      23. Bud with 2 or 3 visible scales; leaf scars half-circular to circular.
         24. Bundle scars in an ellipse; twigs bristly hairy; buds dome-shaped, to 4 mm long _____ **Broussonetia.**
         24. Bundle scars scattered or obscure; twigs glabrous or short-hairy; buds ovoid, to 7 mm long.
            25. Buds asymmetrical, with shorter outer scale bulging to one side; twigs zigzag; pith circular in cross-section _____ **Tilia.**
            25. Buds symmetrical; twigs straight; pith stellate in cross-section _____ **Castanea.**
      23. Buds with 4 or more visible scales (difficult to count in small buds of *Celtis*, but these buds strongly appressed against the twig, unlike buds of above genera).
         26. Low shrubs to 2 m tall; inner bark yellowish, the twigs with rank odor; leaf scars in > 1 plane _____ **Rhamnus.**
         26. Small to large trees, or if shrubby, lacking above combination of features; leaf scars in 1 plane (*Maclura* may vary).
            27. Twigs with milky sap; bundle scars clearly > 3, usually in an elliptic arrangement, not in 3 groups; leaf scars nearly circular; fruit multi-seeded.
               28. Spur branches present; fruit resembling a grapefruit in size and shape, with remnants often present in vicinity of plants (unarmed forms of osage orange) _____ **Maclura.**

28. Spur branches absent; fruits resembling a blackberry, but absent in winter _____ **Morus.**
27. Twigs lacking milky sap; bundle scars 3 or in 3 groups or obscure; leaf scars triangular to half-circular; fruit 1-seeded, a nutlet or samara or drupe.
    29. Bark warty in mature individuals, smooth and gray when immature; lateral buds appressed against the twig; fruit fleshy, a drupe _____ **Celtis.**
    29. Bark otherwise (smooth only in *Carpinus*); lateral buds spreading, at least slightly from twig (tip of bud not against twig surface); fruit dry.
        30. Buds attached directly above leaf scar, not tilted at an angle, the bud scales 6–12, spiraling or in > 2 ranks, and faintly to strongly striated; bundle scars 3, not depressed, often obscure; fruit of hard, flattened or ribbed nutlets.
            31. Bark smooth and sinewy; buds angled and faintly striated; nutlets attached to 3-lobed bract____ **Carpinus.**
            31. Bark brownish and shredding into loose strips; buds rounded and strongly striated; nutlets inside a bladdery sac _____ **Ostrya.**
        30. Buds attached to one side of leaf scar, distinctly tilted at an angle away from the axis of the twig, the bud scales 4–8, in 2 rows, smooth or hairy; bundle scars 3–6, depressed in leaf scar and distinct; fruit a samara or a leathery nutlet.
            32. Bark furrowed; buds 4–8 mm long; fruit a samara; medium to tall trees, widespread in uplands and lowlands _____ **Ulmus.**
            32. Bark platy; buds < 4 mm long; fruit a leathery nutlet with numerous projections from fruit wall; small tree chiefly of Coastal Plain swamps _____ **Planera.**

# Section IV. Generic and Species Accounts

🌿/🌲 **Abies** P. Mill. Fir. Family Pinaceae. (*abeo* = rising-one, the ancient Latin name for a tall ship or tree). Evergreen tree to 25 m tall; bark gray and smooth or with resin blisters, becoming scaly with age; needles spirally arranged but spreading in one plane; the needles flat, blunt, to 2.5 cm long, tapered to disc-like base on twig; cones to 10 cm long, the cone scales and bracts deciduous, and only the cone axis persistent into winter. A tea (good for coughs and fevers) can be made from the needles; the resin (available in bark blisters) when used as a salve or lotion has some antiseptic properties. The dried inner bark can be eaten.

! **A. fraseri** (Pursh) Poir. Fraser f. At high elevations, BR of TN, above 4500 ft (1370 m), and associating with *Picea rubens* to form the spruce-fir zone above 5500 ft (1680 m). A Southern Appalachian endemic, absent from KY, Threatened in TN, largely due to the combined effects of acid rain and infestations of balsam woolly adelgid (*Adelgis piceae*). (for John Fraser, 1750–1811, nurseryman of Chelsea, England). Plate 1.

🌿/☠/➤ **Acer** L. Maple. Family Aceraceae or Sapindaceae. (*acris* = sharp). Small to large trees, bark smooth on younger stems, scaly-furrowed when older; twigs slender, the pith continuous; terminal buds present, rounded to acute, 2–many scales; leaf scars opposite, often curved or V-shaped, touching or nearly so, with bundle scars 3 or in 3 groups; stipular scars absent; fruit a double-samara, persistent in some species. Leaf remnants simple and palmately lobed and veined in all species except *A. negundo*, which has compound leaves with 3–5 leaflets. Maple fruits can be eaten in emergency situations—extract the seeds from the winged samaras and cook like peas (but they are not very tasty). Sugar maple is the best species for sap production, and the sap flows best on warm sunny days after freezing nights. The dried leaves of maples, especially those of red maple, may produce toxic effects if ingested. The inner bark is a good source of fiber for cordage.

1. Terminal bud scales 2, valvate, the buds often constricted and stalked at the base; shrubs or small trees to 10 m tall.
    2. Buds and twigs pubescent; terminal buds to 5 mm long, lacking a distinct stalk; bark unstriped _____ **A. spicatum.**
    2. Buds and twigs glabrous; terminal buds to 12 mm long, distinctly stalked; bark white-striped _____ **A. pensylvanicum.**
1. Terminal buds with 4 or more visible scales, imbricate or valvate, the buds unstalked; trees to 30 m tall.
    3. Terminal buds 6–8 mm long, blunt and ovoid; sap milky; introduced species _____ **A. platanoides.**
    3. Terminal buds to 5 mm long; sap not milky; native species.
        4. Leaf scars meeting and projecting upward into a tooth reaching higher than the tips of the lateral buds; buds blunt and gray-woolly; twigs green; bark ridged and furrowed _____ **A. negundo.**
        4. Leaf scars not completely meeting, or meeting and forming a low tooth; buds not woolly; twigs red to brown to green; bark scaly in older trees (smooth when young).
            5. Buds brown to blackish, with 8 or more visible scales; buds sharp-pointed, rarely in clusters (hard maples).
                6. Small to medium-sized trees of the Coastal Plain and SE TN; bark pale (white, pale brown, to gray), and twigs slender (to 2 mm thick).
                    7. Trunk tall and straight; bark gray and furrowed or platy _____ **A. floridanum.**
                    7. Trunks usually multiple, shrublike; bark whitish, especially on upper trunk and limbs _____ **A. leucoderme.**
                6. Tall trees of the Appalachians and Interior Low Plateaus; bark dark (brown to dark gray), and twigs more robust (> 3 mm thick) _____ **A. saccharum/A. nigrum.**

5. Buds orange to reddish, with 6 or fewer visible scales; buds usually blunt, and often in clusters, especially later in the winter season (soft maples).
    8. Bark with rank odor when scratched; twigs usually brownish _____ **A. saccharinum.**
    8. Bark lacking rank odor; twigs usually reddish or greenish.
        9. Twigs glabrous _____ **A. rubrum.**
        9. Twigs pubescent _____ **A. drummondii.**

**A. drummondii** Hook. & H.J.Arn. ex Nutt. Drummond red m. Lowlands, swamps. Chiefly in CP of KY and TN, with some records from W IP of each state. Infrequent. FACW+, OBL. (probably named in honor of Thomas Drummond, ca. 1790–1835). Plate 2.

**A. floridanum** (Chapm.) Pax. Florida m. Stream banks and moist woods of CP and W IP of KY and TN. Infrequent. Not pictured.

! **A. leucoderme** Small. Chalk m. Bluffs and dry woods of SE TN. Rare. Special Concern in TN. (*leucoderme* = white-skinned). Not pictured.

**A. negundo** L. Boxelder. Woodlands, especially wet woods and along stream banks. Across KY and TN. Abundant. FAC+, FACW. (*negundo* from *nirgundi*, a tree with leaves like box-elder). Plates 3 and 4.

**A. nigrum** F. Michx. Black m. Wet woods, lowlands and lower slopes. Across KY and TN, but most common in calcareous areas in central portions of the states. Infrequent. FACU. Not pictured.

**A. pensylvanicum** L. Striped m. Mixed mesophytic forests. S AP of KY, more common in E TN. FACU, FACU–. Plates 5 and 6.

\* **A. platanoides** L. Norway m. Roadsides and disturbed areas, often planted. Rare escape in IP of KY and TN, introduced from Europe. (*Platanus-oides* = plane-tree-like). Plate 7.

**A. rubrum** L. Red m. Lowland to upland forests. Across KY and TN. Frequent. Two varieties: var. **rubrum**, FAC; and var. **trilobum** Torr. & A.Gray ex K.Koch, FACW, OBL, indistinguishable in winter. Plate 8.

**A. saccharinum** L. Silver or Water m. Swamps, wet woods, along streams. Across KY and TN. Frequent. FACW. (*saccharinum* = sugary). Plate 9.

**A. saccharum** Marshall. Sugar m. Mesic to dry forests. Across KY and TN. Frequent. FACU–. (*saccharum* = sugary). Plates 10 and 11.

! **A. spicatum** Lam. Mountain m. Mesic woods, thickets, chiefly of mid- to high-elevation forests of E TN and N AP of KY, with disjunct populations in IP of both states. Infrequent. Endangered in KY. FACU–, NI. (*spicatus* = with a spicate inflorescence). Plate 12.

**Aesculus** L. Buckeye. Family Hippocastanaceae or Sapindaceae. (*Aesculus* = durmast oak). Medium to tall trees, bark becoming platy on larger specimens; twigs stout, pith thick; terminal buds present, ovoid, sharp-pointed, many-scaled (some twigs tipped by a pair of buds with a fruit scar between them), to 15 mm long; leaf scars opposite, sometimes connected by a ridge, large and shield-shaped or triangular, with bundle scars 3–9 (sometimes in 3 groups); stipular scars absent; fruit a large leathery capsule with large dark-brown seeds (buckeyes). All parts are highly toxic. Seeds are somewhat similar to edible chestnuts (*Castanea* spp.) and hazelnuts (*Corylus* spp.), but ingestion can result in vomiting, paralysis, or death. Note: buckeyes are shiny chocolate brown, rounded on all sides, with a large white scar, and are 2–4 cm wide, whereas chestnuts are flattened on one side and have a pointed tip, and 2 cm or less wide, and hazelnuts also have a large white scar, but are 2 cm or less in width, with a pointed tip on one side, and have a somewhat striped surface. According to folklore, carrying buckeyes in your pocket will cure rheumatism and bring good luck.

1. Capsules spiny; twigs with rank odor when scratched_____ **A. glabra.**
1. Capsules smooth; twigs lacking rank odor.
    2. Trees to 30 m tall; capsule wall > 3 mm thick; distributed chiefly in Appalachian uplands ____**A. flava.**
    2. Shrubs or small trees to 15 m tall; capsule wall 1–3 mm thick; distributed chiefly to south and west of the Appalachians.
        3. Capsules > 4 cm long _____ **A. pavia.**
        3. Capsules < 4 cm long _____ **A. sylvatica.**

**A. flava** Sol. Yellow or Sweet b. Mesophytic forests. Chiefly E KY and E TN, but also scattered through central portions of each state. Frequent. Plates 13 and 14.

**A. glabra** Willd. Ohio b. Dry to mesic woods, usually calcareous. Chiefly IP, both KY and TN. Frequent. FACU+, FACU. (*glabra* = smooth). Plates 15 and 16.

! **A. pavia** L. Red or Scarlet b. Mesic to wet woods. Chiefly CP of both KY and TN. Infrequent. Threatened in KY. FAC. (*pavia* = from Pavia, Italy). Plate 17.

**A. sylvatica** Bartr. Painted b. Wooded slopes and stream banks. AP and VR of TN. FAC. Frequent. Similar to *A. pavia* in winter, but this species is limited to E TN. (*sylvatica* = of the woods). Not pictured.

**Ailanthus** Desf. Family Simaroubaceae. (*aylanto* = tree-of-heaven). Tree to 20 m, with smooth or pale-striped bark; twigs thick, with conspicuous lenticels and large brown pith; terminal buds absent; lateral buds small, about 3 mm long, rounded, 4-scaled, and partly surrounded by leaf scar; leaf scars alternate, large and shield-shaped, with about 9 bundle scars; stipular scars lacking; fruit a samara in large clusters. The scraped bark and foliage exude an unpleasant odor (like burned peanut butter).

** **A. altissima** (Mill.) Swingle. Tree-of-heaven. Disturbed places, woodlands. Across KY and TN. Infrequent, cultivated from eastern Asia and widely escaping. Severe Threat in KY and TN. (*altus* = the tallest). Plates 18 and 19.

**Akebia** Decne. Family Lardizabalaceae. (*Akebia* = Japanese name *akebi*). Woody vines, climbing by twining; twigs brown, lenticels conspicuous; buds about 12-scaled, spreading away from sides of twigs; leaves evergreen, palmately compound; leaf scars elevated on projections from sides of twigs, the bundle scars 3 or more or obscure; fruit a fleshy pod with many black seeds.

** **A. quinata** (Houttuyn) Decne. Five-leaf akebia. Disturbed places, parks. Rare escape, but locally abundant at a few sites in N IP of KY, a native of Asia. Significant Threat in KY. (*quinata* = 5-parted). Plate 20.

**Albizia** Durazz. Family Fabaceae. (for Filippo Albizzi, Italian naturalist). Tree to 12 m tall, with short trunk and spreading branches, bark thin and grayish or greenish, often horizontally striped; twigs zigzag, sometimes lined, with lenticels conspicuous, with green pea odor when broken; terminal buds absent; lateral buds small, about 2 mm long, with visible scales 2 or 3, the buds often superposed; leaf scars alternate, somewhat 3-lobed, with bundle scars in 3 groups; hard stipular projections present, or falling and leaving a wedge-shaped scar; fruit a flat, papery legume about 15 cm long. The legumes are likely toxic.

\*\* **A. julibrissin** Durazz. Mimosa. Thickets, fencerows, woodland margins. Across KY and TN. Frequent, naturalized from tropical Asia. Significant Threat in KY and Severe Threat in TN. (*julibrissin* = silken). Plates 21 and 22.

☤ **Alnus** Ehrh. Alder. Family Betulaceae. (*Alnus* = ancient Latin name for the alder). Shrubs or trees; twigs slender, with pith continuous; terminal buds absent; lateral buds few-scaled, ovoid, about 7 mm long; leaf scars alternate, more or less in one plane, raised, with 3 bundle scars (sometimes broken); stipular scars usually evident; twigs typically with both woody female catkins of previous season (resembling small pine cones) about 15 mm long, and slender male catkins present in winter; fruit a small nutlet. Remnant leaves are simple, ovate, pinnately veined, and toothed. The bark was used for a variety of treatments by Native Americans, including wound healing (chewed bark) and poison-ivy rashes (boiled inner bark).

1. Trees; remnant leaves notched or flat at apex _____**A. glutinosa.**
1. Shrubs; remnant leaves roundly tapering to a point.
    2. Buds and male catkins sessile on twig; twigs usually glabrous; buds shiny and dark; plants of restricted high Appalachian distribution in E TN _____**A. viridis.**
    2. Buds and male catkins stalked; twigs often pubescent; buds dull and rough; plants of wetlands, common across KY and TN_____**A. serrulata.**

\* **A. glutinosa** (L.) Gaertn. Black a. Strip mines, disturbed woodlands. FACW–, NI. Infrequent, a Eurasian species occasionally escaped from cultivation in KY and TN, often planted on strip mines for reclamation. (*gluten* = sticky, with glue). Not pictured.

**A. serrulata** (Aiton) Willd. Smooth a. Wet woodlands and swamps, often along streams. Across KY and TN. Frequent. OBL, FACW+. (*serratus* = edged with small teeth). Plates 23, 24, and 25.

! **A. viridis** (Villars) Lam. & DC. subsp. **crispa** (Aiton) Turrill. Mountain a. Balds. BR of TN (Roan Mountain, Carter County). FAC. Special Concern in TN. (*viridis* = youthful, fresh green). Plate 26.

**Amelanchier** Medik. Serviceberry. Family Rosaceae. (*Amelanchier* = provincial name for snowy-mespilus). Shrubs or small trees with smooth gray bark, becoming furrowed in older specimens; twigs slender, zigzag, pith continuous; terminal buds present, lanceolate and pointed, to 15 mm long, with about 6 twisted scales, these imbricate and ciliate-margined; leaf scars alternate, in one plane, crescent- or U-shaped, with 3 bundle scars; stipular scars absent; fruit a small pome, but absent in winter. Remnant leaves are simple, ovate, pinnately veined, and toothed. The fruits, like miniature apples, mature by late summer, and can be eaten fresh or used in any recipe calling for fruit, although the several species vary considerably in flavor. The tree-sized species provide good branches for constructing arrow shafts. The foliage may be mildly toxic.

1. Plants with a single or few erect stems, usually a small tree, to 15 m tall.
    2. Remnant foliage smooth_____**A. laevis.**
    2. Remnant foliage pubescent _____**A. arborea.**
1. Plants growing in clumps or stoloniferous, to 8 m tall.
    3. Remnant leaves with teeth < 5 per cm_____**A. sanguinea.**
    3. Remnant leaves with teeth > 6 per cm_____**A. canadensis.**

**A. arborea** (F.Michx.) Fernald. Common s. Dry to wet woodlands of slopes, plateaus, and bottoms. Across KY and TN. Frequent in eastern regions, less frequent westward. FAC–, FACU. (*arbor* = tree like). Plate 27.

**A. canadensis** (L.) Medik. Canadian s. Upland woods. E TN only. FAC. Not pictured.

**A. laevis** Wiegand. Smooth s. Mountain forests and woodlands. Chiefly E TN, with single record from E KY. Infrequent. (*laevis* = smooth). Not pictured.

**! A. sanguinea** (Pursh) DC. Red s. Riverbanks and open woods. AP and VR of TN. Threatened in TN. (*sanguinea* = blood-red). Not pictured.

**Amorpha** L. False indigo. Family Fabaceae. (*Amorpha* = deformed-one). Shrubs, often forming thickets; terminal buds absent; buds blunt and scaly, about 3 mm long, somewhat sunken in some species, often superposed; leaf scars alternate, with bundle scars 3 or indistinct; stipular projections often present; fruit a small legume with 1 or 2 seeds, the fruit wall sometimes glandular, and fruits often present in winter.

1. Calyx and legume eglandular or sparsely glandular; upper side of legume nearly straight _____ **A. nitens.**
1. Calyx and legume conspicuously glandular; upper side of legume curved.
    2. Calyx glabrous, the lobes rounded to obsolete _____ **A. glabra.**
    2. Calyx pubescent, the lobes evident and acute _____ **A. fruticosa.**

**A. fruticosa** L. Tall f. i. Wet woods and along streams. Across KY and TN. Infrequent. FACW. (*fruticis* = of shrub-like habit). Plates 28 and 29.

**A. glabra** Desf. ex Poir. Mountain f. i. Woodlands and stream banks. C and E TN. (*glabrous* = smooth). Not pictured.

**A. nitens** F.E.Boynton. Shining f. i. Wet woods and along streams. W KY and C TN. (*niteo, nitere* = with a polished surface, shining). Plates 30 and 31.

☠ **Ampelopsis** Michx. Family Vitaceae. (*Ampelopsis* = vine-resembling). Woody vines with smooth or furrowed bark (not shreddy as in *Vitis* spp.), the pith white and continuous (or diaphragmed in older twigs), and climbing by forked tendrils, these attached opposite a leaf scar; terminal buds absent; leaf scars alternate, with about a dozen bundle scars; stipular scars long and narrow; fruit a black or blue berry, but not very palatable, and may be toxic.

1. Twigs hairy_____ **A. brevipedunculata.**
1. Twigs glabrous.
    2. Buds visible; leaf scars rounded; fruit shiny black_____ **A. arborea.**
    2. Buds sunken; leaf scars notched at apex; fruit blue_____ **A. cordata.**

**A. arborea** (L.) Koehne. Pepper-vine. Wet woods and swamp forests of alluvial bottoms, stream banks, and disturbed sites. Chiefly in western portions of KY and TN, occasional eastward, possibly from plants escaping from cultivation. Infrequent. FACW, FAC+. (*arbor* = tree like). Plates 32 and 33.

* **A. brevipedunculata** (Maxim.) Trautv. Porcelain-berry. Disturbed places. Few records from N and C KY. Infrequent, native of Asia, and commonly cultivated, often persistent, and infrequently escaping. (*brevis-pedunculus* = short-peduncled). Not pictured.

**A. cordata** Michx. Heartleaf pepper-vine, raccoon-grape. Wet woods and swamp forests of alluvial bottoms, stream banks, and disturbed places. Across KY and TN. Infrequent. FAC+. (*cordata* = heart-shaped). Plate 34.

☠ **Aralia** L. Family Araliaceae. (*Aralia* = origin uncertain, possibly French-Canadian, *aralie*). Small tree to 10 m tall, the twigs thick and covered with prickles, bark brownish or grayish and furrowed in older specimens; terminal bud present, imbricate-scaly, large and conical, about 1 cm long; leaf scars narrow, nearly encircling the twig, with > 20 bundle scars; stipular scars absent; fruit a black berry, numerous in broad umbellate clusters, but usually dispersed by winter; calyx remnant on opposite side from fruit stalk. Contact with the resin in the bark and roots may cause skin irritations in some, and the berries should not be consumed, at least not in large quantities (the seeds, in particular, may cause digestive problems).

**A. spinosa** L. Devil's walkingstick. Open mesic to dry forests. Across KY and TN. Frequent. FAC. (*spinae* = spiny). Plates 35 and 36.

**Aristolochia** L. Family Aristolochiaceae. (*Aristolochia* = best childbirth). Woody vines climbing by twining, bark brownish, smooth with vertical splits; twigs greenish, somewhat swollen at nodes; terminal buds absent; buds with 2 silky scales, superposed and sunken in a silky pad nearly encircled by the leaf scar; leaf scars alternate, very narrow, with 3 bundle scars; stipular scars absent; fruit a cylindric capsule, dehiscing to release flat, triangular seeds.

1. Twigs glabrous _____ **A. macrophylla.**
1. Twigs softly pubescent _____ **A. tomentosa.**

**A. macrophylla** Lam. Dutchman's pipe. Mixed mesophytic forest. E KY and E TN, with occasional outliers in IP. Infrequent, occasionally cultivated. (*macrophylla* = with large leaves). [*Isotrema macrophylla* (Lam.) C.F.Reed]. Plate 37.

**A. tomentosa** Sims. Woolly Dutchman's pipe. Swamps and wet woods. Chiefly CP and IP of KY and TN. Infrequent, occasionally cultivated. FAC. (*tomentum* = thickly matted with hairs). [*Isotrema tomentosa* (Sims) C.F.Reed.]. Plates 38 and 39.

**Aronia** Medik. Chokeberry. Family Rosaceae. (a derivative from *Aria*, a name used by Theophrastus for a whitebeam). Shrubs to 3 m tall, with bark smooth or with angular splits; terminal buds present, slender and pointed, 8–12 mm long, scales few and imbricate; leaf scars alternate, crescent- or U-shaped, with 3 bundle scars; stipular scars absent; fruit a small red to purple pome, with calyx remnant on opposite side from fruit stalk, persistent into winter, and edible; twigs may have cherry-like odor. Remnant leaves are simple, ovate, pinnately veined, and toothed.

1. Twigs and buds glabrous or nearly so _____ **A. melanocarpa.**
1. Twigs and buds conspicuously pubescent _____ **A. arbutifolia.**

**A. arbutifolia** (L.)Pers. Red c. Wet woods and swamps, of plateaus and bottoms. Across S KY and across TN, but less common in C TN and W TN. Infrequent. FACW. (*Arbutus-folium* = with *Arbutus*-like leaves). [*Photinia pyrifolia* (Lam.) K.R.Robertson & Phipps]. *A. prunifolia* (Marsh.) Rehd., purple chokeberry, is a supposed hybrid of *A. arbutifolia* and *A. melanocarpa*, and has pubescent twigs like *A. arbutifolia*, but fruits purplish rather than reddish. The purple chokeberry is also known as *Photinia floribunda* (Lindl.) K.R.Robertson & Phipps. Plate 40.

**A. melanocarpa** (Michx.) Elliott. Black c. Wet woods and swamps. Across KY and TN, but rare in central and western portions. Infrequent. FAC. (*melanocarpa* = with very dark or black fruits). [*Photinia melanocarpa* (Michx.) K.R.Robertson & Phipps]. Plates 41 and 42.

**Arundinaria** Michx. Family Poaceae. (*Arundo* = cane or reedlike). Woody grass, sometimes surpassing 7 m tall; stems jointed at nodes; leaves usually present, alternate, with parallel venation; fruit a grain (caryopsis) in spikelets, but plants rarely flowering. Wild cane was widely used by Native Americans for such purposes as the construction of baskets, dwellings, and the making of weapons (i.e., arrows, bows, and blowguns).

1. Branches with 2 or more compressed basal internodes; leaves usually deciduous _____ **A. appalachiana.**
1. Branches with 1 or no compressed basal internodes; leaves persistent _____ **A. gigantea.**

**A. appalachiana** Triplett, Weakley, & L.G.Clark. Hill cane. Woods and thickets. SE TN. Plate 43.

**A. gigantea** (Walter) Muhl. Giant cane. Wet to mesic woods and thickets. Across KY and TN. FACW, FACW. (*gigantea* = very large). Plate 44.

🌿/🍃/➤ **Asimina** Adans. Family Annonaceae. (from the French Canadian name *aisminier*). Small tree to 13 m tall, with bark brownish-gray, smooth or with rough patches, the scraped bark with a sharp, unpleasant petroleum-like, skunklike, or peppery odor; twigs brown, the pith continuous in new growth but diaphragmed in older twigs; terminal buds present, fuzzy-red-brown, naked, to 12 mm long, oblong, flower buds rounded; leaf scars alternate, in one plane, half-round or V-shaped, raised, with 5–7 bundle scars; stipular scars absent; fruit a large fleshy berry, cylindric, to 10 cm long, maturing in early fall. Remnant leaves are simple, obovate, pinnately veined, and entire. The fruits are often eaten raw or made into a pudding or bread. Some people may experience allergic reactions from handling the foliage or gastrointestinal problems from eating the fruit; seeds may contain toxins (known to induce vomiting). Extracts from the stems, leaves, and seeds have been used for medicinal purposes, and may prove to have anti-cancer properties. Fiber from the inner bark can be used to make rope and string.

**A. triloba** (L.) Dunal. Pawpaw. Mesic slopes and bottomland forests. Across KY and TN. Frequent. FACU+, FAC. (*triloba* = three-lobed). Plate 45.

🌿 **Baccharis** L. Family Asteraceae. (*Baccharis* = an ancient Greek name, perhaps "the spicy smell of roots"). Shrubs to 3 m tall, with bark greenish and lined; twigs angled; terminal buds usually absent; buds small and scaly, globose, often obscured by resin; leaf scars alternate, angular, raised, the bundle scars indistinct; stipular scars absent; fruits in heads, a modified achene (cypsela) capped by a cluster of white bristles. At least some leaves are persistent into winter, being simple and few-toothed, but are potentially toxic and should not be eaten.

**B. halimifolia** L. Sea-myrtle. Disturbed areas. In C TN and W TN and recently discovered in W KY (Bruton & Estes 2009). FACW, FAC. (*halimifolia* = gray-leaved). Plates 46 and 47.

🍃 **Berberis** L. Barberry. Family Berberidaceae. (*barbaris* = barberry). Spiny shrubs, with yellow inner bark; twigs often with clustered scaly buds; leaf scars alternate, small, with bundle scars 3 or indistinct; bud clusters and leaf scars subtended by the spines, and stipular scars absent; fruit a red berry. The fruits are edible, but may have poor taste, and are mostly cooked into juice or jelly. Overingestion may cause digestive problems.

1. Spines unbranched _____ **B. thunbergii.**
1. Spines branched _____ **B. canadensis.**

! **B. canadensis** Mill. Canada b. Rocky woodlands. E KY and E TN. Endangered in KY and Special Concern in TN. An alternate host for black stem rust of wheat. Plate 48.

** **B. thunbergii** DC. Japanese b. Woodlands. Across KY and TN. Infrequent, escaped from cultivation, a native of Japan. Significant Threat in KY and TN. FACU, UPL. (*thunbergii* = for Carl Thunberg, 1743–1822). Plate 49.

🌿 **Berchemia** Neck. Family Rhamnaceae. (for M. Berchem, seventeenth-century French botanist). Woody vine climbing by twining, with smooth bark; twigs tough and very flexible; buds and leaf scars small and indistinct; leaf scars alternate, strongly raised, with bundle scars 1 to 3 or indistinct; stipular scars absent; fruit a dark blue drupe. All parts considered toxic. Remnant leaves are simple, pinnately veined, with low teeth.

! **B. scandens** (Hill) K.Koch. Alabama supple-jack. Swamps. CP and IP of KY and TN, infrequent eastward, and much less common in KY. Threatened in KY. FACW. (*scandens* = climbing). Plates 50 and 51.

**Betula** L. Birch. Family Betulaceae. (*Betula* = pitch, from Pliny). Trees, the bark often striped with horizontal lenticels, the color varying as described below in key to species; twigs slender, with conspicuous lenticels and aromatic in some species; terminal buds absent; buds ovoid and sharp-pointed, usually with 2 or 3 imbricate scales, often on short stubby spur branches; leaf scars alternate, more or less in one plane, oval- to crescent-shaped, with 3 bundle scars; stipular scars present; fruiting catkins 2–4 cm long. Remnant leaves are simple, ovate to elliptic or deltoid, pinnately veined and toothed or double-toothed. Twigs and inner bark of some species can provide emergency foods and teas, especially that of sweet and yellow birch. The inner bark can be eaten fresh, shredded and boiled, or dried and ground into flour. The loose outer bark in some species (e.g., river birch) can be used as a fire starter. Birch branches are among the best for making arrow shafts.

1. Bark pinkish or brownish to dark brown and very scaly in older trees, peeling away in paper-like sheets; buds with a bent or hooked tip, the scales hairy-margined; common trees of acid lowlands, and frequently cultivated _____**B. nigra.**
1. Bark otherwise, lacking above combination of color and scaliness; buds not bent or hooked, the scales varying in pubescence; trees of uplands in the Appalachians.
    2. Twigs with wintergreen odor when scratched; bark gray to yellow; common species of the mid- to upper elevations of the Appalachians.
        3. Bark dark gray, smooth; twigs, buds, and catkins glabrous_____**B. lenta.**
        3. Bark yellowish, often peeling or papery; twigs, buds, or catkins pubescent _____ **B. alleghaniensis.**
    2. Twigs lacking wintergreen odor; bark pinkish-white, often in papery strips; very rare trees of high-elevation conifer forests of E TN_____**B. cordifolia.**

**B. alleghaniensis** Britton. Yellow b. Mixed mesophytic forests. E KY and E TN, with disjunct populations in W IP of KY. Infrequent. FAC, FACU+. (*alleghaniensis* = from the Alleghany mountains). Plates 52, 53, 54, and 55.

! **B. cordifolia** Regel. Heart-leaf b. Known only from mountain slope in Sevier County, TN. Endangered in TN. (*cordifolia* = heart-shaped leaf). Plate 56.

**B. lenta** L. Sweet b. Mixed mesophytic forests. E KY and E TN. Infrequent. FACU. (*lentus* = sticky, tough). Plates 57, 58, and 59.

**B. nigra** L. River b. Wet woods and swamps, often along streams. Across KY and TN, but rare in central portions of both states. Frequent. FACW. Plates 60 and 61.

**Bignonia** L. Family Bignoniaceae. (for Abbe Jean Paul Bignon, 1662–1743, librarian to Louis XIV). Woody vine climbing by tendrils; twigs slender, often angled, the pith pale and porous, appearing cross-like in older twigs; terminal bud absent; buds with a few thickened scales, < 5 mm long; leaves scars opposite, narrow, with single bundle scar, the leaf scars connected by a ridge; stipular scars absent; fruit a large flat capsule to 15 cm long by 2.5 cm wide, with winged seeds; the compound leaves, consisting of 2 leaflets and a terminal forked tendril, often persist into the winter, or at least a petiolar stalk persistent.

**B. capreolata** L. Crossvine. Dry to mesic forest. Across KY and TN. Frequent. FAC+. (*capreolus* = sprawling, twining, tendrilled). Plates 62, 63, and 64.

➤ **Broussonetia** L'Hér. Family Moraceae. (for Pierre Marie August Broussonet, 1761–1807, Professor of Botany at Montpellier, France). Small tree, with bark gray to yellow, smooth or shallowly ridged; twigs bristly pubescent, with milky sap (sap may be sparse in winter); pith large, with thin green diaphragms at the nodes; terminal buds absent; buds conical, usually 2-scaled, the scales ridged; leaf scars alternate or opposite, the scars rounded with > 3 bundle scars, often in an ellipse; stipular scars narrow and conspicuous; plants rarely producing viable fruits in the KY–TN region, the trees persisting from plantings or spreading by root sprouts. Remnant leaves are simple, unlobed to palmately lobed, palmately veined, and toothed. The bark is a good source of fiber.

\*\* **B. papyrifera** (L.) Vent. Paper-mulberry. Old home sites, along railroads, disturbed places. Across KY and TN. Infrequent, an occasional escape from plantings, native to Asia. Significant Threat in TN. (*papyrus-fero* = paper-bearing). Plates 65 and 66.

**Brunnichia** Banks. Family Polygonaceae. (for M.T. Brunnich, eighteenth-century Danish naturalist). Woody vines climbing by tendrils, these emerging from same side of twig as the leaf scar; twigs lined, greenish, with encircling lines at the nodes; buds inconspicuous; leaf scars alternate, bundle traces obscure; fruit a winged nutlet in twisted, saclike covering, these often in large clusters.

**B. ovata** (Walter) Shinners. Buckwheat vine. Shoreline and swamp thickets. CP and W IP of KY and TN. Infrequent. FACW. (*ovatus* = egg-shaped). Plates 67 and 68.

**Buckleya** Torrey. Family Santalaceae or Thesiaceae. (for S.B. Buckley, 1809–1884, early American botanist). Shrub to 3 m tall, parasitic on woody plants (chiefly eastern hemlock, *Tsuga canadensis*), the bark thin and smooth; twigs greenish-gray, the lenticels prominent; terminal buds absent, often with paired end buds; buds narrow-lanceolate, about 7 mm long, imbricate scaly, the scales acuminate-tipped; leaf scars opposite or subopposite, half-round, not connected by a line, with single bundle scar; stipular scars absent; fruit a greenish drupe, inedible, and absent in winter.

! **B. distichophylla** (Nutt.) Torr. Pirate-bush. Dry thin woods. BR of TN. Threatened in TN. (*distichophylla* = one-planed leaves). Plate 69.

**Buddleja** L. Family Buddlejaceae. (for Adam Buddle, 1660–1715, English vicar and botanist). Shrubs to 3 m tall, the bark thin and peeling; twigs ridged; terminal buds absent; buds few-scaled, white-woolly with stellate hairs; leaf scars opposite, with 3 bundle scars, the scars connected by a ridge-like line, often with clusters of dried leaves clustered in axils, the leaves simple and white-woolly below; small 2-valved capsules often present in winter.

\*\* **B. davidii** Franch. Butterfly-bush. Disturbed sites, native of China. Escaped across TN and KY, and expected to become more invasive (an Alert species in TN). (*davidii* = for l'Abbé Armand David, collector of Chinese plants in mid-1800s). Plate 70.

**Callicarpa** L. Family Verbenaceae or Lamiaceae. (*callicarpus* = beautiful fruit). Shrub to 2.5 m tall, with a rank odor in inner bark; twigs pubescent with stellate hairs; terminal buds present; buds slender and pointed, naked, grayish; leaf scars opposite, not connected by a line, half-circular, with single bundle scar; stipular scars absent; fruit a pinkish to purplish drupe, the calyx remnant at base of fruit, often persistent in axillary clusters. Remnant leaves are simple, ovate, pinnately veined, and toothed.

**C. americana** L. American beautyberry, French-mulberry. Mesic woods and thickets, sometimes cultivated. Across TN except northern-most sections. FACU–. Plates 71 and 72.

☠ **Calycanthus** L. Family Calycanthaceae. (*Calycanthus* = calyx-flower). Shrub to 3 m tall, aromatic; twigs dark brown, with conspicuous lenticels, somewhat flattened and swollen at nodes; terminal bud absent; buds naked, dark, about 3 mm long, leaf scars opposite, raised and not connected by a line, horseshoe-shaped, with 3 bundle scars; stipular scars absent; fruit of achenes enclosed in a leathery bag-like pseudocarp, often persisting into winter. Remnant leaves are simple, ovate to lanceolate, pinnately veined, and toothed. The fruits and seeds contain chemicals similar to strychnine, and they may produce toxic effects if eaten in quantity.

! **C. floridus** L. Eastern sweetshrub, Carolina allspice. Mesic forests. E KY and TN, scattered westward. Threatened in KY. Two varieties: var. **floridus**, with pubescent stems; and var. **glaucus** (Willd.) Torr. & A.Gray, with smooth stems. (*floridus* = florid, ornate, flowery; *glaucus* = bluish-waxy). FACU, FACU+. Plates 73 and 74.

**Calycocarpum** Nutt. Family Menispermaceae. (*Calycocarpum* = cup fruit). Woody vines climbing by twining; twigs gray-green, finely lined; terminal buds absent; buds small and inconspicuous; leaf scars alternate, large and oval, with 3-many bundle scars; stipular scars absent; fruit an inedible black drupe to 2.5 cm long, the seed cup-like. Other species in this family are poisonous, but there is little information on the toxicity of this plant.

**C. lyonii** (Pursh) A.Gray. Lyon's cupseed. Alluvial bottoms and stream banks. CP and IP of KY and TN. Infrequent. FACW, FACW–. (*lyonii* = for John Lyon, ca. 1765–1814, introducer of American plants). Plates 75 and 76.

☠ **Campsis** Lour. Family Bignoniaceae. (*Campsis* = curvature). Woody vine climbing by aerial roots; twigs gray to brown, speckled; terminal buds absent; buds valvate-scaled, about 3 mm long; leaf scars opposite, shield-shaped, connected by lines, with single or broken bundle scar; stipular scars absent; fruit a capsule to 12 cm long, with winged seeds. Contact with leaves and flowers cause skin irritations for some people.

**C. radicans** (L.) Seem. ex Bureau. Trumpet creeper. Dry to mesic, open and forested sites. Across KY and TN. Frequent. FAC. (*radicans* = with rooting stems or leaves). Plates 77 and 78.

**Carpinus** L. Family Betulaceae. (*Carpinus* = hornbeam). Small tree to 8 m tall, the bark smooth, gray, fluted with "muscular" appearance; twigs slender, zigzag, and shiny; terminal buds absent; buds ovoid, to 9 mm long, imbricate scaly with 8 or more scales, the scales dark with paler margins; leaf scars alternate, in one plane, crescent-shaped, with 3 bundle scars; stipular scars present; fruit a nutlet attached to a leafy 3-lobed green bract. Remnant leaves are simple, ovate to lanceolate, pinnately veined, and double-toothed.

**C. caroliniana** Walter. American hornbeam, ironwood. Stream banks, forested wetlands, and other mesic woodlands. Across KY and TN. Frequent. FAC. Plate 79.

**Carya** Nutt. Hickory. Family Juglandaceae. (*Carya* = Dion's daughter, who was changed into a walnut tree). Trees, the bark often diagnostic (see key to species below); terminal buds present, ovoid, 5–20 mm long, with scales either valvate (in pecan hickories) or imbricate (in true hickories), pubescent in some species; lateral buds sometimes superposed; leaf scars alternate, large and often 3-lobed, with bundle scars many and scattered or in 3 groups; stipular scars absent; fruit hard and nutlike, with smooth surface, enclosed in a hard husk, either completely or partially dehiscent. All hickories have edible nuts but some are bitter, as in *C. cordiformis* and *C. aquatica*. Others have thick shells or small kernels, such as *C. pallida*, *C. tomentosa* and

*C. glabra*. Shellbark, *C. laciniosa*, has a sweet kernel but very thick shell. The two species that produces the best nut for eating, with a sweet nutmeat and thin shell, are pecan (*C. illinoinensis*) and shagbark (*C. ovata*). The inner bark of hickories is a good source of fiber, and strong bows can be constructed from the branches and young saplings.

1. Buds with valvate scales; husk winged (pecan hickories).
    2. Buds yellowish; twigs smooth or scurfy-yellow; bark light to dark gray, usually tightly furrowed and ridged or slightly flaky; widespread in uplands and lowlands_____ **C. cordiformis.**
    2. Buds gray to red-brown; twigs hairy; bark brown to gray with thick, loose scales; lowland species of W KY and TN.
        3. Buds gray-brown; nuts elliptic in length and circular in cross-section_____**C. illinoinensis.**
        3. Buds red-brown; nuts oval in length and flattened in cross-section _____**C. aquatica.**
1. Buds with imbricate scales; husk unwinged (true hickories).
    4. Bark shaggy; nuts compressed and strongly 4-angled; husk to 15 mm thick.
        5. Twigs brownish; husk to 4 cm long and wide _____ **C. ovata.**
        5. Twigs orangish; husk to 6 cm long and 5 cm wide_____**C. laciniosa.**
    4. Bark tight; nuts rounded or only slightly 4-angled; husk to 5 mm thick.
        6. Twigs > 6 mm thick; buds 10–20 mm long, the outer bud scales deciduous and buds appearing white through winter _____**C. tomentosa.**
        6. Twigs 3–6 mm thick; buds 6–12 mm long, the outer scales persistent through winter.
            7. Twigs glabrous or nearly so _____**C. glabra.**
            7. Twigs scaly and often hirsute-pubescent _____ **C. pallida.**

! **C. aquatica** (F.Michx.) Nutt. Water h. Swamps, wet woods. CP and W IP of KY and TN. Threatened in KY. OBL. (*aquaticus* = living in water). Plates 80 and 81.

**C. cordiformis** (Wangenh.) K.Koch. Bitternut h. Bottomlands to dry slopes. Across KY and TN. Frequent. FACU+, FAC. (*cordiformis* = heart-shaped). Plates 82 and 83.

**C. glabra** (Mill.) Sweet. Pignut h. Bottomland to ridgetop woodlands. Across KY and TN. Frequent. FACU–, FACU. Some authors recognize *C. ovalis* (Wangenh.) Sarg., the red hickory, as distinct from pignut hickory. Red hickory can be distinguished by more scaly bark, usually 7 leaflets, and fruit husk that splits to the base, unlike typical pignut with ridged bark, usually 5 leaflets, and fruit husk that splits only at the apex. (*glabra* = smooth, without hairs). Plates 84 and 85.

**C. illinoinensis** (Wangenh.) K.Koch. Pecan h. Rich bottomlands along rivers. CP and W IP, but escaped across both KY and TN. Infrequent. FACU, FAC+. Plates 86, 87, and 88.

**C. laciniosa** (F.Michx.) Loudon. Shellbark h. Rich slopes and bottoms. Across KY and TN, but less frequent eastward. Infrequent. FAC, FACW–. (*lacinia, laciniae* = jagged, fringed). Plates 89 and 90.

**C. ovata** (Mill.) K.Koch. Shagbark h. Dry to mesic slopes. Across KY and TN. Frequent. FACU–, FACU. (*ovatus* = egg-shaped, with broad end lowermost). Two varieties: var. **ovata**, with buds tan to dark brown and densely hairy; and var. **australis** (Ashe) Little, with more slender twigs and buds red-brown, mostly smooth. This variety is treated by some authors as a distinct species, *C. carolinae-septentrionalis* (Ashe) Engl. & Graebn., of dry limestone woods, across TN and S KY. Infrequent. (*australis* = southern; *carolinae-septentrionalis* = from North Carolina). Plates 91, 92, and 93.

**C. pallida** (Ashe) Engl. & Graebn. Sand h. Dry gravelly soils, oak-hickory woods. Across KY and TN, but rare in both the Shawnee Hills region of KY and in C TN. Infrequent. (*pallidus* = greenish, somewhat pale). Plates 94 and 95.

**C. tomentosa** (Poir.) Nutt. Mockernut, white h. Dry slopes and uplands. Across KY and TN. Frequent. [*C. alba* (L.) Nutt. ex Elliott]. (*tomentum* = thickly matted with hairs). Plates 96 and 97.

⚘ **Castanea** Mill. Chestnut. Family Fagaceae. (*Castanea* = the sweet chestnut). Shrubs or trees, with fissured bark; twigs with pith angled in cross-section; terminal buds absent; buds ovoid, with 2 or 3 scales; leaf scars alternate, more or less in one plane, half-circular, the bundle scars > 3 and scattered; stipular scars present; fruit an edible nut, to 2 cm long, enclosed in sharp-spiny husk. Remnant leaves are simple, ovate to elliptic, pinnately veined, and toothed. American chestnut now rarely produces viable fruits, and the dwarf chestnuts are rare to infrequent in the region. If a thicket of dwarf chestnuts can be located, collect the nuts in early autumn, split the shells and roast over heat; for storage, keep nuts in cool dry place, or shell and blanch the nutmeat, and then dry. Do not confuse chestnuts with poisonous buckeye seeds (see under *Aesculus*)!

1. Twigs and and buds glabrous; formerly a large tree, now existing mostly as root sprouts to 7 m tall, and only rarely taller_____**C. dentata.**
1. Twigs and buds pubescent; plant a shrub or small tree.
    2. Spiny husks > 4 cm long and thick, breaking into 4 sections, with 3 nuts; solitary trees to 13 m tall _____**C. mollissima.**
    2. Spiny husks to 2.5 cm long and thick, breaking into 2 sections, with a single nut; shrubs, rarely over 7 m tall, often thicket-forming_____**C. pumila.**

! **C. dentata** (Marshall) Borkh. American c. Dry and mesic acidic woodlands; persisting as stump sprouts after the chestnut blight epidemic, few reaching tree size. Across KY and TN, but absent from BG of KY. Endangered in KY and Special Concern in TN. (*dentatus* = having teeth, with outward pointing teeth). Plates 98, 99, and 100.

* **C. mollissima** Blume. Chinese c. Lawns and old home sites. Across TN and C and E KY; native of Asia, persistent after cultivation, or occasionally escaping, and often confused with American chestnut. (*mollis* = the softest). Plates 101 and 102.

! **C. pumila** (L.) Mill. Allegheny chinkapin, dwarf c. Dry rocky woodlands. E KY and TN, less common in E IP. Threatened in KY. (*pumilio, pumilionis* = very small, dwarf). Plates 103 and 104.

**Catalpa** Scop. Catalpa. Family Bignoniaceae. (*Catalpa* = from East Indian vernacular name, katuhlpa). Trees to 25 m tall, with rounded crowns, the bark scaly or fissured; twigs thick, swollen at nodes; terminal buds absent; buds globose, about 2 mm long, and scaly; leaf scars whorled, rounded and crater-like, with many bundle scars in a circle; stipular scars absent; fruit an elongate capsule with winged seeds. Remnant leaves are cordate, 15–30 cm long, and entire. The leaves are often fed upon by the "catalpa worm," the larval form of the American hawk moth, and these larvae often sought after as fish bait.

1. Capsules > 10 mm thick and often > 40 cm long; seed bodies rounded to the tuft of hairs ____**C. speciosa.**
1. Capsules < 10 mm thick and < 40 cm long; seed bodies tapered to the tuft of hairs ____ **C. bignonioides.**

* **C. bignonioides** Walter. Southern c. Woodlands. Across KY and TN. Infrequent, native to S U.S., occasionally escaping from cultivated plants. UPL, FAC–. (*Bignonia-oides* = bignonia-like). Plates 105 and 106.

**C. speciosa** Warder ex Engelm. Northern c. Bottomland forests and along streams. Across KY and TN. Infrequent, native to W and C portions of the states, but often cultivated and escaping. FAC, FACU. (*speciosa* = showy). Plates 107 and 108.

🌿 **Ceanothus** L. Family Rhamnaceae. (*Ceanothus* = Greek name used by Theophrastus). Shrubs to 1 m tall, with reddish roots; twigs greenish or reddish; terminal buds usually absent; buds ovoid, small, and hairy; leaf scars alternate, half-rounded, with single bundle scar; stipular scars present, or stipules persistent; fruit a small 3-lobed capsule, in terminal panicles, the fruits or their saucer-like bases persistent into winter. A tea, made from the dried rootbark, was used as a sedative by Native Americans.

1. Flowering stalks attached laterally in leaf axils; capsule lobes with ridges _____ **C. americanus.**
1. Flowering stalks terminating main branches; capsule lobes lacking ridges _____ **C. herbaceus.**

**C. americanus** L. New Jersey tea. Dry woods, barrens. Across KY and TN. Infrequent. Plates 109 and 110.

! **C. herbaceus** Raf. Prairie-redroot. Rocky stream banks. Known from only a few records in SE KY and E TN. Threatened in KY. (*herbaceus* = not woody, low growing). Not pictured.

🌿 **Celastrus** L. Bittersweet. Family Celastraceae. (*Celastrus* = Theophrastus' name for an evergreen tree retaining fruit over winter). Woody vines climbing by twining, with orangish roots; terminal buds absent; buds ovoid, 6-scaled, often projecting at a right angle from the twig; leaf scars alternate, half-rounded, with single bundle scar; stipular scars tiny; fruit a orange-yellow capsule, splitting to reveal scarlet-colored arils enclosing the seeds, the fruits persistent into winter. All parts are considered to be toxic, and the fruits, although attractive, should not be ingested.

1. Bud scales rounded and nonspiny; fruit clusters in terminal clusters _____ **C. scandens.**
1. Bud scales keeled and spine-tipped; fruit clusters axillary _____ **C. orbiculatus.**

\*\* **C. orbiculatus** Thunb. Oriental b. Disturbed places. Across KY and TN, infrequent, but locally abundant and increasing in occurrences, naturalized from Asia. Severe Threat in KY and TN. UPL, NI. (*orbis* = disc-shaped, circular in outline). Plates 111 and 112.

**C. scandens** L. American b. Dry to mesic woods, roadsides, fencerows. Across KY and TN, but very uncommon in W IP of KY and C TN. Infrequent. FACU. (*scandens* = climbing). Plate 113.

**Celtis** L. Hackberry. Family Ulmaceae or Cannabaceae. (*Celtis* = ancient Greek name for a tree with sweet fruit). Small to large trees, the bark smooth when young and becoming conspicuously warty when older; twigs slender, zigzag, with pith chambered or continuous in internodes (usually chambered at nodes); terminal buds absent; buds ovoid, to 5 mm long, with 4 or 5 scales visible, appressed against the twig; leaf scars alternate, conspicuously raised, in one plane, half-round, with bundle scars 3 or obscure; stipular scars narrow; fruit a drupe, orangish to red-black. Remnant leaves are simple, ovate to lanceolate, 3-veined from the base, and toothed or entire. Hackberries can be eaten, but the seeds are large and there is not much flesh; taste can vary, but sometimes with a sweet pulp.

1. Shrubs, with leaning or arching trunks; buds 1–2 mm long; drupes purple to black, 5–8 mm long _____ **C. tenuifolia.**
1. Trees, with vertical trunk; buds 3–4 mm long; drupes variable.
    2. Drupe 7–13 mm long, purplish, with a persistent peg-like style _____ **C. occidentalis.**
    2. Drupe 5–9 mm long, dark orange to red, lacking a persistent style _____ **C. laevigata.**

**C. laevigata** Willd. Sugarberry. Dry to wet woods, stream banks. Across KY and TN. FACW. (*laevis* = smooth). Not pictured.

**C. occidentalis** L. Northern h. Dry to wet woodlands, on rocky limestone slopes, and in bottoms. Across KY and TN. Frequent. FACU. (*occidentalis* = of the West). Plate 114.

**C. tenuifolia** Nutt. Dwarf h. Dry woods, limestone slopes and ridges. Across KY and TN, rare in CP and chiefly IP in KY. Infrequent. (*tenuifolia* = slender leaved). Plate 115.

☠ **Cephalanthus** L. Family Rubiaceae. (*Cephalanthus* = head flower). Shrub to 3 m tall, the bark with scaly ridges; twigs with conspicuous lenticels, swollen at nodes, often dying back in winter; terminal buds absent; lateral buds conical-scaly, small and sunken; leaf scars opposite or whorled, rounded with a single U-shaped bundle scar; stipules present and conspicuous, or deciduous and leaving a connecting line between leaf scars; fruit a ball-like cluster of achenes. Remnant leaves are simple, ovate to lanceolate, pinnately veined, and entire. The foliage is likely toxic.

**C. occidentalis** L. Common buttonbush. Shallow ponds, open wetlands, stream and lake margins. Across KY and TN. Frequent. OBL. (*occidentalis* = of the West). Plate 116.

✿/⚕/☠ **Cercis** L. Redbud. Family Fabaceae. (ancient Greek name for a European species of *Cercis* called the Judas-tree). Tree to 10 m tall, the bark with thin scaly ridges; twigs zigzag, purplish with conspicuous white lenticels; terminal buds absent; lateral buds ovoid, to 3 mm long, sometimes superposed, the flower buds stalked, imbricate scaly; leaf scars alternate, in one plane, half-rounded, raised and with two lines extending down from base, with a fringe of red hairs across upper margin, with 3 bundle scars; stipular scars absent; fruit a flat, dark brown legume, 6–10 cm long. Remnant leaves are simple, cordate, palmately veined, and entire. The buds and flowers can be eaten, and the bark has astringent qualities, but there are also reports of some toxic effects.

**C. canadensis** L. Eastern r. Mesic woodlands. Across KY and TN. Frequent. FACU–, FACU. Plates 117 and 118.

✿ **Chaenomeles** Lindl. Family Rosaceae. (*Chaenomeles* = gaping apple). Shrub to 3 m tall, with thorns above the leaf scars; terminal buds absent, the twig usually tipped by a thorn; buds few-scaled, rounded, in pairs at base of thorns; leaf scars alternate, half-round, bundle scars 3 or broken; stipular scars usually not evident; fruit a sour pome, apple-like in size and appearance, produced in the fall, and edible when cooked.

* **C. speciosa** (Sweet) Nakai. Flowering quince. Occasionally escaping, often persistent after cultivation. Across KY, rare in TN. Native of China. (*speciosa* = showy). Plate 119.

**Chimaphila** Pursh. Family Pyrolaceae or Ericaceae. (*Chimaphila* = winter love, wintergreen). Evergreen subshrub to 15 cm tall; leaves alternate, simple, often clustered and appearing opposite or whorled, to 6 cm long and 2 cm wide, lanceolate, sharply serrate; fruit a 5-parted capsule in clusters on long naked peduncles.

**C. maculata** (L.) Pursh. Spotted-wintergreen. Dry to mesic forests, hardwood and pine-hardwood. Chiefly in E KY and E TN, less common westward, and absent from lowlands of CP. Frequent. (*maculata* = spotted, blotched). Plate 120.

⚕ **Chionanthus** L. Family Oleaceae. (*Chionanthus* = snow flower). Shrub or small tree to 6 m tall, with furrowed and scaly bark; twigs stout, swollen at nodes and somewhat angled, pubescent when young; terminal buds present, buds low and rounded, about 5 mm long, with

3 or 4 pairs of awn-tipped scales; leaf scars opposite, not connected by lines, almost circular, with single bundle scar; stipular scars absent; fruit an inedible purplish drupe to 2 cm long, the calyx remnant at base of fruit. Remnant leaves are simple, elliptic, pinnately veined, and entire. The bark (boiled or crushed) was used by Native Americans for skin inflammations and wound healing.

**C. virginicus** L. Fringetree. Dry to mesic forests, bluffs and cliffs. Chiefly E KY and E TN, less common westward in IP. Infrequent. FAC+, FACU. Plate 121.

**Cladrastis** Raf. Family Fabaceae. (*Cladrastis* = brittle-branched). Tree to 20 m tall with smooth gray bark, the wood and inner bark yellow; twigs zigzag, dark brown, speckled with light-colored lenticels; terminal buds absent; buds conical, to 5 mm long, naked with silky-brown hairs, the buds nearly encircled by leaf scar; leaf scars alternate, 2-ranked, narrow and horseshoe-shaped, with 5 bundle scars; stipular scars absent; fruit a papery, pale legume to 10 cm long, often with stamen remnants present at base of fruit, these persistent into winter.

**C. kentukea** (Dum.Cours.) Rudd. Kentucky yellowwood. Mesic to dry calcareous woods. KY and TN, chiefly IP and AP, scattered to very rare in other regions. Infrequent and local. Plates 122 and 123.

⚘ **Clematis** L. Clematis. Family Ranunculaceae. (*Clematis* = Greek name for several climbing plants). Slightly woody vines climbing by twisting petioles; twigs and pith with 5 or more angles, the twigs sometimes pubescent; buds small, ovoid, with pubescent scales; leaves opposite, the blades or petioles evergreen or persistent; fruit of achenes with elongate feathery styles, these often persistent. The foliage is likely toxic, but was used medicinally by Native Americans for a wide variety of ailments.

1. Leaves present in winter, the leaflets 3–5, entire_____**C. terniflora.**
1. Leaves absent in winter (petioles may be persistent).
   2. Achenes 35 or fewer per flower; remnant leaves with 5 or more leaflets_____ **C. catesbyana.**
   2. Achenes 40 or more per flower; remnant leaves with 3 leaflets _____ **C. virginiana.**

! **C. catesbyana** Pursh. Catesby's virgin's-bower. Open limestone sites in IP, very rare in KY, more common in C TN. Historical in KY. (*catesbyana* = for Mark Catesby, 1674–1749, author of the first published account on the flora and fauna of North America—*Natural History of Carolina, Florida, and the Bahama Islands*). Plate 124.

** **C. terniflora** DC. Sweet autum c. Native of Asia, cultivated and escaping, chiefly in IP and AP of KY and TN. Lesser Threat in TN. FACU–, FAC–. [*C. dioscoreifolia* H.Lév. & Vaniot.; *C. paniculata* Thunb.]. (*terni-florum* = flowers in threes). Plate 125.

**C. virginiana** L. Virgin's-bower. Mesic woods. Across KY and TN. Frequent. FAC, FAC+. Plates 126 and 127.

**Clethra** L. Pepperbush. Family Clethraceae. (*Clethra* = ancient Greek name for alder). Shrubs or small trees to 6 m, with reddish, shreddy bark; twigs with stellate pubescence; terminal buds present, conical, naked and silvery-silky, to 5 mm long; leaf scars alternate, triangular, clustered toward twig tip, with single bundle scar; stipular scars absent; fruit an ovoid capsule, to 5 mm long, in terminal racemes. Remnant leaves are simple, elliptic to lanceolate, pinnately veined, and toothed.

1. Shrub 3–6 m tall; plants of southern Appalachians _____ **C. acuminata.**
1. Shrub < 2.5 m tall; plants of southern U.S. _____ **C. alnifolia.**

**C. acuminata** Michx. Mountain p. Mesic to dry forests. AP of KY and TN and BR of TN. Infrequent. (*acuminate* = with long, narrow and pointed tip). Plates 128 and 129.

! **Clethra alnifolia** L. Coastal p. Wet woods. E IP of TN, known only from Coffee County, TN. Endangered in TN. FAC+, FACW. (*alnifolia* = alder-leaved). Plate 130.

☠ **Cocculus** DC. Family Menispermaceae. (*Cocculus* = small berry). Woody vine climbing by twining; twigs green; buds naked and hairy, about 3 mm long; leaf scars alternate, elliptic, bundle scars ≥ 3; stipular scars absent; fruit a poisonous red drupe with crescent-shaped seed. Remnant leaves are simple, palmately veined, unlobed or lobed, with entire margins.

**C. carolinus** (L.) DC. Coralbeads. Upland and lowland woods and thickets, often in sandy or calcareous soils, and in disturbed sites. Frequent across KY and TN, but apparently very rare in AP of KY. FAC. Plates 131 and 132.

**Comptonia** L'Hér. Family Myricaceae. (for Henry Compton, 1632–1713, Bishop of Oxford, then Bishop of London). Aromatic shrub to 1 m tall; twigs pubescent, resin-dotted; terminal buds absent; buds rounded, about 4-scaled; leaf scars alternate, raised, triangular, with 3 bundle scars; stipular scars tiny; fruit a nut surrounded by bristly bracts; cylindric, many-scaled catkin buds usually present in winter, as are a few persistent leaves, these simple and deeply pinnately lobed, resin-dotted and aromatic.

! **C. peregrina** (L.) J.M.Coult. Sweet-fern. River cobble bars. Disjunct from northern U.S. range in 3 counties in AP of KY and TN. Endangered in KY and TN. (*peregrinus* = strange, foreign, exotic). Plates 133 and 134.

**Conradina** A.Gray. Family Lamiaceae. (*Conradina* = for an unidentified lady). Minty-aromatic evergreen shrub, the lower branches sprawling and rooting, with some branches rising to 20 cm tall; leaves opposite, simple, linear, to 2.5 cm long, the margins inrolled, in clusters; fruit of 4 nutlets enclosed in calyx tube.

! **C. verticillata** Jennison. Cumberland rosemary. Sandy river banks. Endemic to 6 counties in AP of KY and TN. Federally Threatened, Endangered in KY, and Threatened in TN. FACW+, FACW–. (*verticillus* = having whorls). Plate 135.

**Cornus** L. Dogwood. Family Cornaceae. (*cornu, cornus* = horn). Shrubs or tree to 15 m tall; twigs slender, appearing "jointed" in opposite-leaved taxa; terminal buds present, ovoid to rounded, few-scaled, usually appearing valvate, often pubescent; leaf scars opposite in all but one species, the leaf scars in opposite-leaved species connected by lines, with bundle scars 3 in all species; stipular scars absent; fruit a red, white, or blue drupe, the calyx remnant at apex of fruit. Remnant leaves are simple, elliptic, lateral veins arching toward leaf apex, entire. The fruits vary in edibility, and may be toxic. Chewing the bark may help alleviate headaches, and a poultice of the bark has been used for treating wounds. The hard wood of the trunk and larger branches is useful for constructing handles for tools, and the smaller branches can be used for making arrows.

1. Leaf scars alternate_____ **C. alternifolia.**
1. Leaf scars opposite.
    2. Buds dissimilar, the flower buds 4-scaled and biscuit-shaped, the vegetative buds 2-scaled, pointed; plant a small tree with blocky bark; drupe red_____ **C. florida.**
    2. Buds similar and pointed; multi-stemmed shrubs with smooth or pebbly bark; drupe white to blue.

3. Pith dark brown, especially in second-year twigs _____ **C. amomum.**
3. Pith white or tan, especially in second-year twigs.
    4. Twigs distinctly pubescent _____ **C. drummondii.**
    4. Twigs glabrous or nearly so _____ **C. foemina/C. sericea.**

**C. alternifolia** L.f. Alternate-Leaved d. Mesic forests. Chiefly in E KY and TN, less common in E IP. Infrequent. (*alternus-folium* = with alternate leaves). Plate 136.

**C. amomum** Mill. Silky d. Wet woods, stream banks. Across KY and TN. Frequent. FACW, FACW+. Frequent. (*amomum* = purifier). Two subspecies: subsp. **amomum**, and subsp. **obliqua** (Raf.) J.S. Wilson, indistinguishable in winter, and the latter chiefly distributed in C and E KY. Plate 137.

**C. drummondii** C.A.Mey. Rough-leaved d. Swamps to upland woods, thickets, and roadsides. CP and IP of KY and TN. Frequent. FAC. (*drummondii* = for one of several Mr. Drummonds, probably Thomas Drummond, ca. 1790–1835). Plate 138.

**C. florida** L. Flowering d. Dry to wet woods. Across KY and TN. Frequent. FACU–, FACU. (*floridus* = florid, ornate, free-flowering, flowery). Plates 139 and 140.

**C. foemina** Mill. Stiff d. Wet woods and swamps. Across KY and TN, but much less common eastward. Infrequent. FAC, FACW–. Two subspecies: subsp. **foemina**, with twigs maroon, inflorescence flat-topped, fruit blue, frequent in CP of both states (*C. stricta* Lam.); and subsp. **racemosa** (Lam.) J.S.Wilson, with twigs tan, inflorescence pyramidal, fruits white, rare in IP and AP of KY. (*femina, feminae* = feminine). Plates 141 and 142.

**C. sericea** L. Red osier d. Known from a few upland sites in IP and AP of KY. FACW+ in Region 1, but KY records mostly from dry sites. Compared to *C. foemina*, this species has redder twigs and leaves 5–7 veined. (*sericea* = silky). Not pictured.

**Corylus** L. Hazelnut. Family Betulaceae. (*Corylus* = helmet). Shrubs to 3 m tall; twigs zigzag, and bearing catkins in winter; terminal buds absent; buds ovoid, with 5 or 6 scales, to 4 mm long; leaf scars alternate, 2-ranked, triangular, with 3–many bundle scars; stipular scars present; fruit a large edible nut, to 1.5 cm long, enclosed in a husk (involucre). Remnant leaves are simple, ovate-cordate, pinnately veined, and irregularly or double-toothed. Nuts mature by early autumn, and can be eaten fresh, and used in any recipe calling for nuts. The nuts have a high oil content and are an excellent survival food. The inner bark is a good source for fiber, and the larger branches or young saplings are good for constructing bows.

1. Twigs with red-glandular pubescence; mature involucres to 3 cm long, the apex wide and jagged; widespread across KY and TN _____ **C. americana.**
1. Twigs lacking glands; mature involucres to 7 cm long, the apex constricted into an elongate beak; restricted to mountains of E TN _____ **C. cornuta.**

**C. americana** Walter. American h. Dry to mesic woodlands, thickets, disturbed sites. Across KY and TN. Frequent. FACU–, FACU. Plates 143, 144, and 145.

**C. cornuta** Marshall. Beaked h. Thickets and woodland borders, often at high elevations. E TN. (*cornuta* = horned). Plates 146, 147, and 148.

**Cotinus** Mill. Family Anacardiaceae. (*Cotinus* = ancient Greek name for a wild olive). Small tree to 15 m tall, the bark gray-scaly, and the sap sticky with pinelike or menthol odor; twigs glaucous and purplish-brown, strongly lenticellate, with brown pith; terminal buds present; buds small, about 6 mm long, the scales imbricate, sharply black-tipped; leaf scars alternate, half-rounded and with folds or wrinkles, with 3 to many bundle scars; stipular scars absent; fruit a kidney-shaped drupe on hairy stalks, the grayish-purplish hairs on the large multi-branched inflorescence providing the "smoky" appearance to the fruiting display. Remnant leaves are simple, elliptic, pinnately veined, and entire.

! **C. obovatus** Raf. American smoke-tree. Limestone woodlands. S AP of TN, on border with Alabama. Special Concern in TN. (*obovatus* = egg-shaped in outline, with the more narrow end lowermost). A similar species, *C. coggygria* Scop., is commonly planted. Plates 149 and 150.

**Crataegus** L. Hawthorn. Family Rosaceae. (*Crataegus* = strong). Small trees, mostly < 10 m tall, usually with thorns at the nodes; twigs with short, stubby spur branches, mostly glabrous; terminal buds present; buds ovoid-obtuse, with thick, gummy imbricate scales; leaf scars alternate, crescent-shaped, with 3 bundle scars; stipular scars present but inconspicuous; fruit a pome, to 1.5 cm thick, with calyx remnant at apex of fruit, usually red, green, or yellow. Remnant leaves are simple, mostly pinnately veined, toothed, sometimes lobed. All species produce edible fruits, which vary in taste, and may persist into winter; the fruits can be eaten fresh, made into jelly or tea, or can be dried for later use. Hawthorn fruits may be one of our best heart-healthy wild foods; they have long been used in Europe and China to improve circulation and heal the heart.

Hawthorn species are very difficult to distinguish in winter conditions. Below is a list of known occurrences in KY and TN. See Jones (2005) and Wofford and Chester (2002) for keys to the taxa in summer condition.

**C. berberifolia** Torr. & A.Gray. Barberry h. Swamps. CP of KY. Rare. FACU, FAC. (*berberifolia* = barbarry-leaved). Not pictured.

**C. calpodendron** (Ehrh.) Medik. Pear h. Upland woods, stream banks. KY and TN, IP, and AP. Infrequent. (*calpodendron* = urn-tree, referring to shape of fruit). Not pictured.

**C. coccinea** L. Scarlet h. AP. Rare. IP and AP of KY. (*coccineus* = crimson, scarlet). Not pictured.

**C. collina** Chapman. Hill h. Uplands. Rare in IP in KY. (*C. punctata* Jacq.). (*collina* = growing on hills). Not pictured.

**C. crus-galli** L. Cockspur h. Fields, woodlands, and thickets. Across KY and TN. Infrequent. FACU, FAC–. (*crus-galli* = cock's spur). Not pictured.

**C. disperma** Ashe. Spreading h. IP and AP of TN. (*dispersa* = scattered). Not pictured.

! **C. harbisonii** Beadle. Harbison's h. Limestone hills and thickets. CP and IP of TN. Endangered in TN. (after T.G. Harbison, 1862–1936). Plate 151.

**C. intricata** Lange. Entangled h. Upland woods and fields. C and E portions of both KY and TN. Infrequent. [*C. boyntonii* Beadle, *C. rubella*]. (*intricare, intrico* = entangled). Not pictured.

**C. macrosperma** Ashe. Fan-leaf h. Uplands and mesic woods. Across KY and TN, but rare in CP. Infrequent. [*C. flabellata* (Spach.) G.Kirchn.]. (*macrosperma* = large-seeded). Not pictured.

**C. marshallii** Eggl. Parsley h. Woodlands. Across S TN, primarily a Coastal Plain species of SE U.S. FACU+, FAC. (after Humphrey Marshall, 1722–1801, U.S. botanist). Not pictured.

**C. mollis** (Torr. & A.Gray) Scheele. Downy h. Woodland edges and thickets. Across KY but restricted to IP of TN. Infrequent. FACU, FAC. (*mollis* = softly hairy, soft). Not pictured.

**C. phaenopyrum** (L.f.) Medik. Washington h. Low woods and fields. W and C KY and TN. Infrequent, also widely cultivated and occasionally escaping. FAC, FAC–. (*phaenopyrum* = with shining grains). Plate 152.

**C. pruinosa** (H.L.Wendl.) K.Koch. Frosted h. Upland and lowland woods. Across TN but restricted to E IP of KY. Infrequent. (*prunia* = powdered, with a hoary bloom as though frosted-over). Not pictured.

**C. punctata** Jacq. Dotted h. Upland mesic woods. C and E TN, including high-elevation sites. Infrequent. (*pungo, punctum* = with a pock-marked surface, spotted). Not pictured.

**C. spathulata** Michx. Littlehip h. Low to upland woods. IP of KY and TN, and in AP/RV of TN. Very rare in KY to infrequent in TN. FAC. (*spatha* = shaped like a spoon). Not pictured.

**C. uniflora** Münchh. One-flower h. Dry uplands, rocky woods. S and E TN, SE KY. Infrequent. (*uni-florum* = one flowered). Not pictured.

**C. viridis** L. Green h. Wet to mesic woods. Chiefy CP and W IP of KY and TN, occasional eastward in TN. Infrequent. FACW. (*viridis* = youthful, fresh green). Plate 153.

☙ **Cytisus** L. Family Fabaceae. (*Cytisus* = the Greek name for a cloverlike plant). Small shrub to 2 m tall, the upper branches wandlike and dying back in winter; twigs heavily ridged and bright green; buds ovoid, about 3 mm long, scaly; leaf scars alternate, raised, tiny, with single bundle scar; stipules minute and persistent; fruit a flat legume to 5 cm long. The leaves are compound with 3 leaflets, and may be present in winter. The plant has toxins, but these are concentrated mostly in the flowers and fruits.

\* **C. scoparius** (L.) Link. Scotch broom. Disturbed places, adventive from Europe. TN only, scattered in eastern half of state. (*scoparium* = broomlike). Plate 154.

**Decodon** J.F.Gmel. Family Lythraceae. (*Decodon* = ten teeth). Arching shrub usually in shallow water, woody stem base with spongy bark, the stems elongating and arching to 3 m above the water, rooting at stem tips, the upper stems sometimes dying back in winter; twigs green, buds reddish, sunken, and inconspicuous; leaf scars opposite or subopposite, lacking a connecting line, with single bundle scar; stipular scars absent; fruit a capsule in axillary clusters. Remnant leaves are simple, lanceolate to elliptic, pinnately veined, with smooth margins.

**D. verticillatus** (L.) Elliott. Swamp loosestrife. Bald-cypress swamps, marshes. CP, scattered eastward in KY and TN. Rare. OBL. (*verticillus* = having whorls). Plate 155.

**Decumaria** L. Family Hydrangeaceae. (*decumma* = 10-parted). Woody vine climbing by aerial roots; twigs zigzag, brown, swollen at nodes, appearing jointed; terminal bud present, acute, reddish; leaf scars opposite, connected by ridge-like lines, V-shaped, with 3 bundle scars; fruit a striped ovoid capsule, the persistent base of the style forming a "cap," the fruits in rounded clusters, and persistent. The leaves may be persistent, and are simple, pinnately veined, and toothed.

**D. barbara** L. Climbing hydrangea. Wet to dry woods, roadsides. CP, S IP, and in SE corner counties of TN. FACW. (*barbara* = foreign, from Barbary). Plates 156 and 157.

**Deutzia** Thunb. Family Hydrangeaceae. (for Johannes van der Deutz, 1743–1788, sheriff of Amsterdam and Karl Pehr Thunberg's patron). Shrub to 3 m tall, with peeling bark; twigs pubescent with stellate hairs, the pith hollow; terminal bud sometimes present; buds ovoid, scaly, the twigs often tipped with a pair of buds; leaf scars opposite and connected, with 3 bundle scars; stipular scars absent; fruit a small flat-topped capsule (often with persistent style) in large terminal panicle, these often persistent and conspicuous in winter. Remnant leaves are simple, lanceolate, pinnately veined, and toothed.

\* **D. scabra** Thunb. Mock-orange. Woods, old home sites. Rare escape, but sometimes locally abundant, scattered sites in KY and TN. Native of E Asia. (*scabra* = with a rough surface to the touch). Plates 158 and 159.

**Diervilla** Mill. Bush honeysuckle. Family Caprifoliaceae or Diervillaceae. (for M. Diéreville, French surgeon and traveler in Canada from 1699–1700). Shrubs to 1.5 m tall, often with shreddy bark; terminal buds absent; buds with loose scales, the lateral buds often appressed to twig; leaf scars opposite, connected by lines, crescent-shaped, with 3 bundle scars; stipular scars absent; fruit a slender capsule, to 1.5 cm long, in terminal clusters. Remnant leaves are simple, lanceolate to ovate, pinnately veined, and toothed.

1. Twigs pubescent _____ **D. rivularis.**
1. Twigs glabrous or with pubescence in lines at nodes.
    2. Twigs angled _____ **D. sessifolia.**
    2. Twigs rounded _____ **D. lonicera.**

! **D. lonicera** Mill. Northern b. h. Dry rocky slopes and thickets. Rare in E TN, outlier in W IP. Threatened in TN. (for Adam Lonitzer, 1528–1586, German physician and botanist). Plate 160.

! **D. rivularis** Gattinger. Mountain b. h. Mesic woods and stream banks. E TN. Threatened in TN. (*rivularis* = of the rivers, waterside). Plates 161 and 162.

**D. sessilifolia** Buckley. Southern b. h. Rocky slopes and stream banks of high elevations. E TN. (*sessilifolia* = unstalked, lacking a petiole). Plates 163 and 164.

**Diospyros** L. Family Ebenaceae. (*Diospyros* = divine food, fruit of the gods). Medium to large trees, the bark often blocky and blackish; twigs glabrous or pubescent, with diaphragmed pith, the fruit stalks persisting on female plants; terminal buds absent; lateral buds with 2 shiny visible scales; leaf scars alternate, ovoid, with single banana-shaped (or lunate) bundle scar; stipular scars absent; fruit a large berry with calyx at base of fruit, falling in late autumn. Remnant leaves are simple, elliptic, pinnately veined, and entire. Fruits mature in late autumn, and should not be collected until after a frost when the flesh becomes softer and sweeter and the skin wrinkled (note: attempts to ingest the berry prior to a frost may result in a drawing up of the mouth and throat and a choking sensation, as the flesh is very astringent). The ripened fruits can be used in breads, puddings, and jam, or frozen for later use.

**D. virginiana** L. Persimmon. Dry to mesic woods. Across KY and TN. Frequent. FAC–, FAC. Plates 165 and 166.

**Dirca** L. Family Thymelaeaceae. (from *Dirce*, in Greek mythology, a fountain in Boeotia). Shrubs with leathery, flexible bark; twigs with jointed nodes, the buds silky-hairy, terminal buds absent; leaf scars alternate, horseshoe-shaped, surrounding the naked bud, with 5 bundle scars; fruit a drupe, varying from green to yellow to red. Remnant leaves are simple, elliptic, pinnately veined, and entire. Both the fruit and the bark contain a irritating resin; ingestion of the fruit can cause digestive problems and contact with the leaves or bark can cause skin irritations similar to those caused by poison-ivy.

**D. palustris** L. Leatherwood. Mesic forests. C and E portions of both KY and TN. Infrequent. FAC, FACU–. (*palus, paludis* = of swampy ground). Plate 167.

**Elaeagnus** L. Family Elaeagnaceae. (*Elaeagnus* = olive-chaste-tree). Shrubs, the twig surfaces covered with silvery or brownish peltate scales, sometimes spiny; terminal buds present;

buds appearing naked or valvate-scaled; stipular scars absent; leaf scars alternate, half-round, with 1 bundle scar; evergreen species with leaves alternate, simple, ovate, the lower surface pale and dotted; fruit drupe-like, red or yellow, actually an achene surrounded by an edible fleshy hypanthium. All our species have been introduced from Asia or Eurasia. They are often planted on strip mines or used for wildlife plantings, and are naturalized across both states. Fruits maturing by late summer and often persist into the winter. A variety of wildlife, especially songbirds such as cedar waxwings, evening grosbeaks, and robins, are attracted to the fruits. Future use of these species should be discouraged because of their invasive nature. One good way to help control this species is to gather and eat the fruits! Some species, especially E. umbellata, are quite tasty and high in vitamin C when ripe (but astringent if unripe), and can be made into jellies and preserves.

1. Leaves evergreen, glossy and leathery _____**E. pungens.**
1. Leaves deciduous, but remnant leaves often persisting.
    2. Remnant leaves silvery or grayish on both sides; taller shrubs or small trees to 7 m tall___**E. angustifolia.**
    2. Remnant leaves silvery only on lower surface, green on upper surface; shrub to 3 m tall.
        3. Twigs and leaf midribs brown-scaly; fruiting stalks > 1 cm long _____ **E. multiflora.**
        3. Twigs and leaf midribs silvery-scaly; fruiting stalks < 1 cm long _____ **E. umbellata.**

\*\* **E. angustifolia** L. Russian-olive. Naturalized from Eurasia, rarely established in IP of KY, and on the Alert list in TN. (*angustifolia* = narrow-leaved). Not pictured.

\* **E. multiflora** Thunb. Asian-olive. Woodland edges and mature forests. Naturalized from Asia, and locally escaping in western portions of C TN (Gorman et al. 2011). (*multiflora* = many flowered). Not pictured.

\*\* **E. pungens** Thunb. Thorny-olive. Naturalized from Asia, rarely escaping in IP of KY, infrequent across TN. Significant Threat in TN. (*pungens* = ending in a sharp point, prickly). Plates 168 and 169.

\*\* **E. umbellata** Thunb. Autumn-olive. Naturalized from E Asia, frequently escaping across KY and TN. Severe Threat in KY and TN. (*umbella, umbellate* = umbelled). Plate 170.

**Epigaea** L. Family Ericaceae. (*epigaeus* = above ground, growing close to ground surface). Evergreen subshrub with prostrate, red-bristly stems and peeling bark; leaves alternate, simple, about 7 cm long by 2.5 cm wide, ovate and entire-ciliate, with cordate base and distinct petioles; fruit a small rounded capsule.

**E. repens** L. Trailing arbutus. Dry to mesic forests. KY and TN, chiefly AP to BR, scattered westward in IP. Infrequent. (*repens* = creeping). Plates 171 and 172.

☠ **Euonymus** L. Family Celastraceae. (*Euonymus* = famed, of good name). Shrubs and woody vines (evergreen in exotic species); twigs green and often 4-angled or 4-winged; terminal buds present, scaly imbricate, acute; leaf scars opposite or subopposite, crescent-shaped, with single bundle scar; stipular scars minute; fruit a purple or red capsule, the seeds with fleshy coverings (arils), these red or orange. Evergreen and deciduous species with leaves simple, ovate-elliptic, pinnately veined, and serrate. The species have a long history of medicinal use in both America and Asia, but the bark, leaves, and fruits are known to contain toxins. Caution is urged because children may be attracted to the colorful seeds.

1. Leaves evergreen, thick and glossy; vines climbing by aerial roots or forming a ground cover ___ **E. fortunei.**
1. Leaves deciduous; erect or trailing shrubs.
    2. Twigs winged with corky growths _____ **E. alatus.**

2. Twigs unwinged.
   3. Twigs 4-angled; stems to 2 m tall; capsule warty.
      4. Stems erect, lacking aerial roots, often 1–2 m tall _____ **E. americanus.**
      4. Stems trailing, with aerial roots, the stems < 5 dm tall_____ **E. obovatus.**
   3. Twigs unangled, or only slightly so, plant a small tree to 6 m tall; capsule smooth _____ **E. atropurpureus.**

\*\* **E. alatus** (Thunb.) Siebold. Winged burning-bush. Disturbed areas, woodlands. Chiefly IP and AP, but spreading across both states. Infrequent, but often locally abundant, introduced from Asia. Severe Threat in KY and Lesser Threat in TN. (*alatus* = wing like, winged). Plates 173 and 174.

**E. americanus** L. Strawberry-bush, Hearts-a-bursting-with-love. Mesic to alluvial woodlands. Across KY and TN. Frequent. FAC, FAC–. Plates 175 and 176.

**E. atropurpureus** Jacq. Eastern wahoo. Woodlands and woodland borders, often in limestone areas. Across KY and TN. Infrequent. FACU, FAC. (*atropurpureus* = dark purple colored). A European species that sometimes escapes in the region, *E. europaeus* L., is similar in twig and fruit, but unlike *E. atropurpureus*, which has smooth bark, leaves pubescent below, and red arils, this exotic species has fissured bark, leaves smooth below, and orange arils. Plates 177 and 178.

\*\* **E. fortunei** (Turcz.) Hand.-Mazz. Climbing euonymus, winter creeper. Disturbed places, woods. Chiefly in C KY and TN. Infrequent, but locally abundant and increasing in occurrence, introduced from Asia. Severe Threat in KY and Lesser Threat in TN. (*fortunei* = for Robert Fortune, 1812–1880, Scottish plant collector for the Royal Horticultural Society in China). (*E. hederaceus* Champ. ex Benth.). Two other evergreen Asian species, *E. kiautschovicus* Loes., and *E. japonicus* Thunb., may occasionally escape in the region, but these species both have a shrubby habit and lack aerial roots on the branches. Plate 179.

**E. obovatus** Nutt. Running strawberry-bush. Mesic forests. AP and IP of KY and TN, more common in BR of TN. Infrequent. (*ob-ovatus* = egg-shaped in outline and with narrow end lowermost, obovate). Plates 180 and 181.

**Fagus** L. Beech. Family Fagaceae. (the Latin name for the beech tree, for the edible seed of beech). Trees to 35 m with smooth gray bark; twigs slender, zigzag; terminal buds present, with 10 or more imbricate scales, about 1 cm long, narrowly elongate and very sharp-pointed, often present at the ends of short spur branches; leaf scars alternate, in one plane, half-round, with 3 or more bundle scars; stipular scars narrow and nearly encircling the twig (these encircling rings may not be apparent when buds are terminating short spur branches); fruit a triangular nut, in pairs, in a small soft-prickly husk, about 2.5 cm long. Leaf remnants are often persistent, and are simple, ovate, pinnately veined, and serrate. Nuts can be gathered after hard frost, then after being dried and oven-heated, the shells can be cracked and the kernels can be eaten raw, ground and used for flour or as a coffee subsitute, or chopped and used in recipes calling for pecans. The nuts have a high oil content and are an excellent survival food. One caveat—there is some evidence that ingestion of large quanties can be toxic, so moderation is advisable.

**F. grandifolia** Ehrh. American b. Mesic forests. Across KY and TN. Frequent. FACU. (*grandisfolium* = with large leaves). Plates 182, 183, and 184.

**Forestiera** Poir. Family Oleaceae. (for Charles Le-Forestier, seventeenth-century French naturalist). Shrubs or small trees, sometimes thorny; twigs hairy, often with sharp spur branches; terminal buds present, small and few-scaled, lateral buds often superposed; leaf scars opposite, not connected by lines, the scars half-round, with single bundle scar; stipular scars absent; fruit a drupe. Remnant leaves simple, lanceolate to elliptic, pinnately veined, and toothed.

1. Plants > 3 m tall; swamplands in western regions of KY and TN; drupes > 1 cm long_____ **F. acuminata.**
1. Plants < 3 m tall; uplands, limestone glades and similar habitats; drupes < 1 cm long _____ **F. ligustrina.**

**F. acuminata** (Michx.) Poir. Swamp-privet. Swamp forests. CP and W IP of KY and TN. Infrequent. OBL. (*acuminate* = with a long, narrow and pointed tip). Plate 185.

! **F. ligustrina** (Michx.) Poir. Upland-privet. Cedar glades, bluffs. IP of KY and TN. FAC (in TN). Threatened in KY. (privetlike, resembling privet, *Ligustrum*). Plate 186.

**Forsythia** Vahl. Family Oleaceae. (for William Forsyth, 1737–1804, superintendent of Kensington Royal Gardens). Loosely branched shrub; twigs smooth, with pale lenticels; pith chambered or excavated in internodes; terminal buds present, with about 6 pairs of scales; leaf scars opposite, not connected by lines, shield-shaped, with 1 bundle scar; stipular scars absent; fruit a small capsule, but rarely fruiting. Remnant leaves are simple, elliptic to lanceolate, pinnately veined, and toothed.

* **F. viridissima** Lindl. Golden bells. Persistent at old home sites, often planted, rarely escaping, often flowering in late winter. Few sites in IP and AP of KY and TN. *F. suspensa* (Thunb.) Vahl., may also occasionally escape. (*viridis* = youthful, fresh green). Plate 187.

**Fothergilla** L. Family Hamamelidaceae. (for John Fothergill, 1712–1780, London physician and botanist, who introduced American witch-hazel to English gardens). Shrub to 2.5 m tall; twigs with stellate pubescence; terminal buds present, the outer scales soon deciduous, and inner buds appear naked and stalked, to 10 mm long; leaf scars alternate, half-round, with 3 bundle scars; stipular scars present; fruit an ovoid capsule. Remnant leaves are simple, ovate to obovate, with stellate hairs, pinnately veined, and coarsely toothed.

! **F. major** (Sims) Lodd. Mountain witch-alder. Cobble bars, Big South Fork area of E TN. Threatened in TN. (*major* = larger, greater). Plates 188 and 189.

**Frangula** Mill. Buckthorn. Family Rhamnaceae. (*frangula* = brittle, fragile). Shrub or small tree to 7 m tall; twigs slender and hairy; terminal buds present, naked, hairy, about 5 mm long; leaf scars alternate, raised, half-round, with 3 bundle scars; stipular scars minute; fruit fleshy, berry-like, red to black, with calyx remnant at base of fruit, inedible. Remnant leaves are simple, elliptic, pinnately veined (very straight-veined), toothed.

**F. caroliniana** (Walter) A.Gray. Carolina b. Dry to mesic woods and thickets, limestone areas. Across KY and TN, but rare in CP and far eastern sections. Frequent. FAC. (*Rhamnus caroliniana* Walter). The Eurasian *F. alnus* Mill., with 2-seeded fruit instead of 3-seeded, may occasionally escape. Plate 190.

**Fraxinus** L. Ash. Family Oleaceae. (*Fraxinus* = ancient Latin name for ash, used by Virgil). Trees to 35 m tall; twigs slender to stout, often compressed at nodes; terminal buds present, acute or blunt, scurfy-surfaced, with 2 or 3 pairs of thick scales; leaf scars opposite, raised, unconnected by lines, half-round to horseshoe-shaped, with many bundle scars in a U- or C-shaped line; stipular scars absent; fruit a terminally-winged samara with a narrow seed body. The inner bark is a good source of fiber for making string and rope, and the branches are useful for making bows.

1. Twigs 4-angled; samaras with flattened seed bodies, scarcely distinguishable from the wing, the wing extending to the base of the seed _____ **F. quadrangulata.**

1. Twigs rounded; samaras with seed bodies rounded and clearly distinguishable from the wing, the wing extending to ¾ or less of the seed body.
    2. Terminal bud with truncate apex; samaras large, to 8 cm long and 13 mm wide; twigs and buds usually pubescent; plants of swamps in western portions of KY and TN, often with swollen trunk bases _____ **F. profunda.**
    2. Terminal buds with acute or obtuse apices; samaras averaging smaller, usually < 5 cm long and 7 mm wide; twigs and buds usually glabrous or nearly so, occasionally pubescent; plants of wetlands and uplands, but rarely in water or with swollen trunk bases.
        3. Buds obtuse; leaf scars usually notched on upper side; samara wing extending to < ½ of seed body _____ **F. americana.**
        3. Buds acute; leaf scars usually straight on upper side; samara wing extending ½ or more of seed body _____ **F. pennsylvanica.**

**F. americana** L. White a. Mesic to dry forests. Across KY and TN. Frequent. FACU. Plates 191 and 192.

**F. pennsylvanica** Marshall. Green a. Alluvial forests, especially along streams. Across KY and TN. Frequent. FACW. Two varieties: var. **pennsylvanica**, the red ash, with twigs and leaves pubescent; and var. **subintegerrima** (Vahl.) Fernald, the green ash, with glabrous parts, are recognized by some authors. Plates 193 and 194.

**F. profunda** (Bush) Bush. Pumpkin a. Swamp forests. KY and TN. Chiefly CP, also in W IP of KY. Infrequent. OBL. [*F. tomentosa* Michx.]. (*profundus* = large, very tall). Plates 195 and 196.

**F. quadrangulata** Michx. Blue a. Upland calcareous forests. KY and TN. Chiefly of C and EC portions of KY and TN, absent from CP. Frequent. (*quadrangularis* = with four angles, quadrangular). Plates 197 and 198.

✤ **Gaultheria** L. Family Ericaceae. (for Dr. Jean Francois Gaulthier, 1708–1756, Swedish-Canadian botanist, author of *Flora Helvetica*). Evergreen rhizomatous subshrub, with aerial branches rising < 15 cm high, the crushed foliage with odor of wintergreen; leaves alternate, simple, to 5 cm long by 2 cm wide, elliptic to obovate, with low teeth toward blunt apex; fruit red and berry-like. A tea can be brewed from the aromatic leaves, and the young leaves and fruits can be eaten raw. Wintergreen tea should be avoided during pregnancy.

**G. procumbens** L. Wintergreen. Dry to mesic forest. KY and TN. E IP to BR. Frequent. FACU. (*procumbo* = lying flat on the ground, creeping forwards, procumbent). Plate 199.

✤ **Gaylussacia** HBK. Huckleberry. Family Ericaceae. (for Joseph Louis Gay-Lussac, 1778–1850, French philosopher and chemist). Shrubs to 1.5 m tall, often colonial; twigs lacking the speckled appearance of some *Vaccinium* species, but often pubescent; terminal buds absent; buds ovoid, to 4 mm long, scales 4–7, imbricate, often with yellow resin dots; leaf scars alternate, half-round to triangular, with single bundle scar; stipular scars absent; fruit berry-like and edible, usually black, with 10 seeds, the calyx remnants at apex of fruit. Leaves in both evergreen and deciduous species simple, elliptic, and pinnately veined. Of these species, *G. baccata* produces the best-tasting berries, and uses are similar to those of blueberries. Both blueberries and huckleberries can be dried for later use.

1. Leaves evergreen and serrulate _____ **G. brachycera.**
1. Leaves deciduous (remnant leaves with margins entire or nearly so).
    2. Twigs with stalked glandular trichomes; plants to 5 dm tall _____ **G. dumosa.**
    2. Twigs lacking stalked glands; plants usually > 5 dm tall.

3. Plants to 1 m tall; buds to 2 mm long, usually with 5–7 visible scales _____ **G. baccata.**
3. Plants 1–2 m tall; buds to 4 mm long, usually with 4 visible scales _____ **G. ursina.**

**G. baccata** (Wangenh.) K.Koch. Black h. Upland oak and pine woodlands. Chiefly in the mountains of KY and TN, AP to BR, and in Land Between The Lakes region of both states. Frequent. FACU. (*baccata* = having berries). Plate 200.

**G. brachycera** (Michx.) A.Gray. Box h. Upland, acid woodlands, often in heath-like communities near cliff edges. S AP of KY and N AP of TN, documented in only 8 counties. Rare, but locally abundant. (*brachycerus* = short-horned). The species is considered to be a relict of the Ice Ages, and some colonies are among the oldest known organisms on Earth, estimated to be between 5K and 13K years old! [*Buxella brachycera* (Michx.) Small]. Plate 201.

! **G. dumosa** (Andr.) Torr. & A.Gray. Dwarf h. Sandy to boggy sites. Eastern IP and western AP of TN, in only 6 counties. FAC. Threatened in TN. (*dumosa* = compact, bushy). Plate 202.

! **G. ursina** (M.A. Curtis) Torr. & A.Gray *ex* A.Gray. Bear h. Wooded slopes. BR of TN and S AP of KY. Infrequent. Threatened in KY. (*ursina* = relating to bears). Plate 203.

**Gelsemium**. Family Loganiaceae or Gelseminaceae. (*Gelsemium* = the Italian for true jasmine). Woody vine climbing by twining; twigs smooth, swollen at nodes, sometimes angled; end buds to 5 mm long, loosely few-scaled, pale; leaves evergreen, simple, lanceolate, entire, and long-acute to acuminate-tipped; leaf scars opposite, connected by a line, crescent-shaped, with single bundle scar; stipular scars absent; fruit a 2-valved, beaked capsule, with winged seeds. All parts considered to be highly poisonous, and bees may use the nectar to produce a toxic honey.

! **G. sempervirens** (L.) A.St-Hil. Yellow jessamine. Woodlands and thickets. Known from 4 counties in southeastern TN (AP and RV). FAC. Special Concern in TN. (*sempervirens* = always green). Plates 204 and 205.

**Gleditsia** L. Family Fabaceae. (for John Gottlieb Gleditsch, 1714–1786, of the Berlin Botanic Garden). Tree to 30 m tall, usually armed with branched thorns on trunk and twigs, the bark with long plates in mature specimens; twigs zigzag, shiny, swollen at nodes, with branched thorns; terminal buds absent; buds small, superposed and sunken in a hairy pit, and concealed by leaf scar and bark; leaf scars alternate, raised, shield- to U-shaped, with bundle scars in 3 groups; stipular scars absent; fruit a flat legume. The pulp of the pods is very sweet and edible, especially before complete ripening (caution—do not mistake for the toxic pods of Kentucky coffeetree, *Gymnocladus dioicus*).

1. Legumes < 5 cm long, seeds 1–3; thorns usually < 15 cm long; remnant leaflets glabrous or nearly so _____ **G. aquatica.**
1. Legumes > 5 cm long, (to 50 cm) long, seeds 4 or more; thorns often > 15 cm long; remnant leaflets pubescent _____ **G. triacanthos.**

! **G. aquatica** Marshall. Water-locust. Bald-cypress swamps. CP of KY and TN, and W IP of KY. Rare. Special Concern in KY. OBL. (*aquaticus* = living in water). Plates 206 and 207.

**G. triacanthos** L. Honey-locust. Dry to mesic woodlands. Across KY and TN. Frequent. FAC–. (*triacanthus* = three-spined). Plates 208 and 209.

**Gymnocladus** Lam. Family Fabaceae. (*Gymnocladus* = naked branch, referring to winter appearance). Trees to 30 m tall, the bark with curling plates when older; twigs very stout, with orange lenticels and large pinkish pith; terminal buds absent; buds small, silky, superposed, and sunken; leaf scars alternate, heart-shaped, with bundle scars 3–5; stipular scars absent or tiny; fruit a heavy, dark-brown legume, to 20 cm long. The fresh seeds and fruit pulp are toxic, but the seeds lose their toxicity when parched (and were used by early settlers as a coffee substitute).

**G. dioicus** (L.) K.Koch. Kentucky coffeetree. Dry to mesic woodlands. Across KY and TN, chiefly IP and CP. Frequent. (*dioicus* = of two houses, referring to the unisexual plants). Plates 210, 211, and 212.

**Halesia** J.Ellis ex L. Family Styracaceae. (for the Reverend Dr. Stephen Hales, 1677–1761, experimentalist and writer on plants). Trees to 30 m tall, with distinctive purplish-scaly bark; twigs pubescent, the young trunks white-striped; terminal buds usually present, conical, to 5 mm long, with 2–4 visible scales, lateral buds often superposed; pith finely chambered or diaphragmed; leaf scars alternate, half-round with a notch, bundle scar single; stipular scars absent; fruit a 4-winged drupe to 4 cm long.

! **H. carolina** L. Silverbell. Mesic ravines associated with Tennessee River in W KY, and historical records from AP of KY; in TN common in VR and BR, and a prominent member of the "cove hardwood" forests of the Smoky Mountains, less common along Tennessee River to W TN. Endangered in KY. FACU, FACU+. [*H. tetraptera* J.Ellis]. Plates 213, 214, and 215.

⚕ **Hamamelis** L. Witch-hazel. Family Hamamelidaceae. (Greek name for a tree that flowers and fruits at the same time). Shrub to 5 m tall with smooth bark; twigs zigzag, stellate pubescent when young; terminal buds present, oblong, to 8 mm long, brownish-yellow and hairy, appearing naked, and stalked; leaf scars alternate, in one plane, half-round to triangular, with 3 bundle scars; stipular scars present, unequal; fruit a 2-beaked woody capsule to 1.5 cm long, with blackish seeds. Remnant leaves are simple, elliptic to obovate, with oblique bases, pinnately veined, and coarsely toothed. The boiled leaves, twigs, and bark of witch-hazel were used for a variety of purposes by the Native Americans, including treatments for respiratory ailments and wound healing, and the species is still widely used in a many modern commercial products.

**H. virginiana** L. American w.-h. Woodlands, stream banks to ridge tops. Across C and E KY and TN. Frequent. FAC–, FACU. Plates 216 and 217.

☠ **Hedera** L. Family Araliaceae. (*Hedera* = the Latin name for ivy). Woody vines with evergreen leaves, trailing and forming a ground cover or climbing by aerial roots; leaves alternate, simple, with palmate venation, young leaves palmately lobed, older ones unlobed, the surfaces glossy-green with veins often whitish; fruit berry-like, black, borne in umbellate clusters, with calyx remnant at apex of fruit, inedible. The fruits have a bitter taste and if ingested can have harmful effects on the digestive, respiratory, and nervous systems.

** **H. helix** L. English ivy. Lawns and old home sites. Across KY and TN. Infrequently escaping, introduced from Europe. Significant Threat in KY and Lesser Threat in TN. (*helix* = winding, the ancient Greek name for twining plants). Plates 218 and 219.

**Hibiscus** L. Family Malvaceae. (*hibiscum* = marsh mallow). Shrub to 3 m tall; twigs pubescent, with terminal buds clustered, but hidden in flat-topped twig apex; leaf scars alternate, often clustered, some raised, bundle scars 4 or indistinct; stipules persistent, long and conspicuous, greatly surpassing the buds; fruit a dehiscent capsule to 2.5 cm long, persistent. Remnant leaves are simple, usually 3-lobed and veined, and coarsely toothed.

\*\* **H. syriacus** L. Rose-of-Sharon. Disturbed places, along railroads. Across KY and TN. Infrequent, introduced from Asia. Lesser Threat in TN. (*syriacus* = from Syria). Plates 220 and 221.

**Hydrangea** L. Hydrangea. Family Hydrangeaceae. (*Hydrangea* = water vessel, referring to the tiny cuplike fruits). Shrubs to 2.5 m tall, the bark often peeling on older stems; twigs slender, soft-wooded, the pith large, with old bud scales persisting at bases of twigs; terminal buds present, globose to ovoid, 4–6 scaled; stipular scars absent; leaf scars opposite, connected by lines, crescent-shaped, with 3–7 bundle scars; fruit a small striped capsule capped by 2 persistent style branches, in flat-topped terminal clusters, persistent. Remnant leaves are simple, ovate, pinnately veined, and toothed (lobed in *H. quercifolia*). Leaves, twigs, and buds have sometimes been used to supplement salad greens, but this is not a good idea—these parts contain toxins that can lead to digestive, respiratory, and nervous system problems.

1. Terminal buds > 1 cm long; bundle scars > 3 _____ **H. quercifolia.**
1. Terminal buds much shorter; bundle scars 3.
   2. Stems pubescent _____**H. radiata.**
   2. Stems glabrous_____ **H. arborescens/H. cinerea.**

**H. arborescens** L. Wild h. Mesic to wet woods. Across KY and TN, from lowlands to highlands. Frequent. FACU. (*arborescens* = becoming or tending to be of treelike dimensions). Plates 222 and 223.

**H. cinerea** Small. Gray h. Indistinguishable from *H. arborescens* in winter, occurring on bluffs and in mesic woods, of KY and TN, most frequent in CP and W IP, less frequent eastward. (*cinis, cinerus* = ash gray). Plates 224 and 225.

**H. quercifolia** W.Bartram. Oak-leaved h. Across S TN, absent from KY. Occasional on bluffs and in dry to mesic woods. (*quercifolius* = having leaves resembling oak leaves). Plates 226 and 227.

**H. radiata** Walter. Silverleaf h. Open woods and roadbanks. TN only, in BR. (*radiata* = radiating outward). Plate 228.

**Hypericum** L. St.-John's-wort. Family Hypericaceae. (*Hypericum* = over an apparition, referring to the practice of hanging the plants over shrines to repel evil spirits). Shrubs with flaking bark; twigs slender, often 2-edged, the pith porous or excavated; leaves sometimes present in winter, especially as clusters in the leaf axils; terminal buds absent; buds scaly and minute, sunken and concealed; leaf scars opposite, connected by a line, triangular, with single bundle scar; stipular scars absent; capsules 4–15 mm long, sometimes persistent. Remnant leaves characterized by translucent or dark dots on the surfaces. Species in the genus, especially the herbaceous species, have a long history of medicinal use for such problems as depression and insomnia.

1. Plants mostly < 0.5 m tall; capsules 1-locular, with 4 sepal remnants at base of capsule; foliage leaves often persistent in winter.
   2. Leaves to 20 mm wide, clasping the stem; fruits with 3–4 styles_____ **H. crux-andreae.**
   2. Leaves to 7 mm wide, tapering to base; fruits with 2 styles.
      3. Plant with erect, usually single stem, often > 30 cm tall_____**H. hypericoides.**
      3. Plants with many decumbent stems, usually < 30 cm tall_____**H. stragulum.**

1. Plants 0.5–2 m tall; capsules 1–3 locular, with 5 sepal remnants at base of capsule; foliage leaves usually absent, but axillary clusters of leaves often persistent.
    4. Stems with axillary clusters of leaves few or lacking; capsule walls incomplete, not dividing the capsule into locules.
        5. Leaves with an articulation at base (note ridge across leaf base at junction with stem); sepals 15–30 mm long; capsules 3 or fewer in sessile cymes _____ **H. frondosum.**
        5. Leaves lacking a joint at base (note that lower midvein is unbroken as it merges with stem); sepals < 10 mm long; capsules many in stalked, branched cymes _____ **H. nudiflorum.**
    4. Stems with axillary clusters of leaves usually numerous; capsule walls meeting, or nearly so, and dividing the capsule into 3–5 locules.
        6. Capsules 7–13 mm long, in clusters of 7 or fewer _____ **H. prolificum.**
        6. Capsules to 6 mm long, in clusters of 8 to many.
            7. Style branches 3 _____ **H. densiflorum.**
            7. Style branches 4 or 5 _____ **H. lobocarpum.**

! **H. crux-andreae** (L.) Crantz. St.-Peter's-wort. Wet meadows. SE KY, E TN, C TN (except CB). Threatened in KY. FACU. (*crux-andreae* = St. Andrew's cross). Plates 229 and 230.

**H. densiflorum** Pursh. Bushy St.-J.-w. Wet meadows and woods. KY (chiefly CP) and TN (chiefly VR, also in C and W TN, except CB). FAC+, FACW–. (*densiflorus* = densely flowered, close-flowered). Plate 231.

**H. frondosum** Michx. Cedarglade St.-J.-w. Limestone glades and bluffs of both states. Rare in IP of KY, more common on Cumberland River bluffs of TN, and occasional in E and W parts of TN. Infrequent. (*frondosus, frondos* = leafy). Plate 232.

**H. hypericoides** (L.) Crantz. St.-Andrew's Cross. Dry to mesic woods. KY (rare in AP and CP), and across TN. FACU, FAC. (*hypericoides* = resembling *Hypericum*). Plate 233.

**H. lobocarpum** Gatt. Fivelobe St.- J.-w. Wet woods and meadows. KY and TN, CP and IP. Rare to infrequent. (*lobocarpus* = having lobed fruits). Plates 234 and 235.

! **H. nudiflorum** Michx. Early St.-J.-w. Marshes, wet woods. Infrequent in AP of TN. Special Concern in TN and historical report for KY. OBL, FACW. (*nudiflorus* = with fully exposed, or naked flowers). Plate 236.

**H. prolificum** L. Shrubby St.-J.-w. Wet to dry woods, often on cliff lines. Across KY and TN, less frequent westward. Infrequent. FACU. (*prolificus* = very fruitful). Plates 237 and 238.

**H. stragulum** W.P.Adams & N.Robson. St.-Andrew's-cross. Dry to mesic, often rocky woods, barrens. Across KY and TN. Infrequent. (*stragulum* = mat-forming). Plate 239.

**Ilex** L. Holly. Family Aquifoliaceae. (*Ilex* = the Latin name for holly oak). Shrubs or small trees to 10 m tall; twigs typically bearing spur branches; terminal buds present, obtuse to ovoid, the lateral buds often superposed, clustered toward apex, bud scales 4–6; leaf scars alternate, crescent-shaped, with single bundle scar; twigs with tiny black stipules or tiny stipular scars; evergreen species with leave simple, ovate to elliptic and spiny-margined; fruit fleshy, berry-like and red, often persisting into winter, the calyx remnant is at base of fruit. American holly is often used for decorations during the winter holidays, and the red fruits can be tempting to children, but they are not edible—ingestion often leads to vomiting and other digestive problems. The fruit and bark have been used to treat a variety of ailments, including fevers, skin problems, and insomnia, but may also cause hallucinations, vomiting, and diarrhea. The straight branches make sturdy arrow shafts.

1. Leaves present, evergreen and spiny _____ **I. opaca.**
1. Leaves absent (remnant leaves serrate to crenate).
    2. Seeds striate-ribbed.
        3. Sepals of fruit entire; plants of central and western regions.
            4. Fruit stalks > 8 mm long_____ **I. longipes.**
            4. Fruit stalks < 8 mm long_____ **I. decidua.**
        3. Sepals ciliate; plants mostly restricted to Appalachian regions _____ **I. ambigua.**
    2. Seeds smooth_____ **I. verticillata.**

**I. ambigua** (Michx.) Torr. Carolina h. Mixed mesophytic forests. AP of KY, and in AP, BR, and few sites westward in TN. Rare. (*ambigo, ambigere* = of ambiguous relationship). Two varieties: var. **ambigua** and var. **montana** (Torr. & A.Gray) H.E.Ahles, distinguished only by leaf features (petioles < 1 cm long and apex more abruptly acuminate in the former). Plate 240.

**I. decidua** Walter. Possum haw. Wet to dry woods. CP and W IP of KY and TN, rare in W IP. Frequent. FACW, FACW–. (*decidua* = deciduous). Plates 241 and 242.

**I. longipes** Chapm. ex Trel. Buckbush. Upland woods anhd thickets. In TN only, CP and southern tier of counties. FAC. (*longipes* = long-stalked). Plate 243.

**I. opaca** Aiton. American h. Mesic forests, wet woods, and swamps. Across KY and TN but less common in limestone regions. Frequent. FACU+. (*opaca* = darkened, dull, shady). Plate 244.

**I. verticillata** (L.) A.Gray. Common winterberry. Swamps and wet woods. Across KY and TN but mostly absent from limestone regions. Infrequent. FACW+, FACW. (*verticillus* = having whorls). Plates 245, 246, and 247.


**Itea** L. Family Grossulariaceae or Iteaceae. (*Itea* = the Greek name for a willow). Much-branched shrub to 3 m tall; twigs with chambered or diaphragmed pith; terminal buds present, appearing naked or valvate, conical, the lateral buds much smaller and globose, with long-pointed scales; leaf scars alternate, half-round, and bundle scars 3; stipules present but scars inconspicuous; fruit a 2-valved, hairy capsule, to 7 mm long, borne in long racemes. Remnant leaves are ovate to elliptic, pinnately veined, and slightly toothed.

**I. virginica** L. Virginia-willow, sweet spire. Alluvial plains and plateaus, especially swamp forests dominated by bald-cypress or red maple-blackgum. Across KY and TN except for far eastern portions of both states, and absent from many counties in BG and CB. OBL, FACW+. Plates 248, 249, and 250.


**Juglans** L. Walnut. Family Juglandaceae. (*Juglans* = Jupiter's-nut, *glans Jovis* in Pliny, referring to legend of Earth-dwelling gods dining on walnuts, while mortal men had to eat acorns; the common name is partly Teutonic in origin, from *wallnuss*, or in Old English, *walh-hnut*). Trees to 30 m tall, with rough or blocky and ridged bark; twigs with chambered pith; terminal buds present, with soft-hairy, valvate scales, to about 1 cm long; lateral often superposed; leaf scars alternate, 3-lobed, with bundle scars in 3 U-shaped groups; stipules absent; fruit hard and nutlike, with sculptured (wrinkled and ridged) surface, in a leathery, indehiscent husk. For both species, the husks should be removed from the nut, and the nut allowed to dry before cracking and extracting the sweet nutmeats. Extracts of the bark, twig, and leaves have been used for a variety of ailments, from healing sores to toothache to ringworm. The husks contain chemicals with antifungal properties, as well as other toxins that have been used to stun fish in sluggish water or ponds. The characteristic odor is due to the chemical compound juglone.

The inner bark is a good source of fiber for making string and rope, and the branches or young saplings are useful for making bows.

1. Bark gray and scaly; terminal buds elongate and pointed and tawny; leaf scars with a fuzzy ridge along upper edge below axillary bud; pith dark brown; fruits ovate-elliptic _____ **J. cinerea.**
1. Bark dark and ridged; terminal buds blunt and gray; leaf scars lacking a fuzzy ridge; pith light-brown; fruits globose _____ **J. nigra.**

! **J. cinerea** L. White w., Butternut. Bottomlands, ravines, moist slopes. Across KY and TN. Threatened in KY and TN. FACU+, FACU–. (*cinis, cinerus* = ash gray). Plates 251 and 252.

**J. nigra** L. Black w. Bottomlands, ravines, upland woods. Across KY and TN. Frequent. FACU. Plates 253 and 254.

**Juniperus** L. Red-cedar, juniper. Family Cupressaceae. (*Juniperus* = the ancient Latin name for Old World junipers). Shrub or tree, with shreddy bark; leaves simple and opposite, in 4 ranks, scale-like (to 4 mm long) or needle-like (to 18 mm long); cone berry-like, fleshy and purplish. The cones are eaten raw, cooked, or used for flavoring (i.e., gin, vodka, and pemmican), but cones are not very tasty and have a high resin content; they can cause stomach troubles or other reactions, and should not be eaten in quantity and should be avoided by pregnant women and those with uterine or kidney problems; the inner bark can be eaten in emergency situations. The boiled foliage and cones have been used medicinally for a variety of ailments (including bronchitis) by the Native Americans. The inner bark is a good source of fiber for making string and rope, and the branches or saplings are useful for making bows.

1. Shrubs, mostly < 2 m tall; needles in whorls of 3; cones axillary _____ **J. communis.**
1. Trees, to 25 m tall; needles opposite; cones terminal _____ **J. virginiana.**

! **J. communis** L. Common j. Dry woods. AP of KY. Threatened in KY. (*communis* = growing in clumps, gregarious, common). Not pictured.

**J. virginiana** L. Eastern r.-c. Dry woods, fields, limestone glades. Across KY and TN. Abundant. FACU, FACU–. Plates 255 and 256.

**Kalmia** L. Family Ericaceae. (for Peter Kalm, 1716–1779, a highly reputed Finnish student of Linnaeus, who spent 3 years, 1748–1751, investigating the botany of northeastern North America). Evergreen shrubs; leaf arrangement various (see below), the leaves simple, leathery, elliptic to lanceolate, to 10 cm long; vegetative buds minute, 2-scaled; flower buds many-scaled, clustered on stalks at twig tips or in leaf axils, persistent into winter; fruit a subglobose capsule, 5-sectioned. All parts are considered to be highly toxic. The flower buds have sometimes been eaten, or a tea made from the leaves, but partaking of these items can have drastic consequences, including paralysis and death. The toxin is also present in flowers, and the consumption of honey made from *Kalmia* flowers can lead to convulsions and paralysis. Similar toxins are found in other genera of the Ericaceae, including *Rhododendron*, *Leucothoe*, and *Menziesia*.

1. Leaves opposite or whorled, pubescent below, mostly < 2 cm wide; shrub < 2 m tall _____ **K. carolina.**
1. Leaves alternate, rarely whorled, glabrous below, mostly > 2.5 cm wide; shrub > 2 m tall ___ **K. latifolia.**

! **K. carolina** Small. Carolina laurel. Bogs. Only known record from Johnson County, TN. Endangered, possibly extirpated. FACW. Plate 257.

**K. latifolia** L. Mountain laurel. Dry to mesic forest. Across KY and TN except for CP of both states. Frequent. FACU. (*latus-folium* = with broad leaves). Plates 258 and 259.

**Kerria** DC. Family Rosaceae. (for William Kerr, who introduced the widely cultivated and double-flowered "*pleniflora*" in the early 1800s). Shrubs to 3 m tall, often with arching stems, spreading by root sprouts; bark and twigs bright green, terminal buds lacking; lateral buds dark red-brown, and collateral buds often present; leaf scars alternate, with 3 bundle trace scars; stipules or stipular scars evident; rarely fruiting. Remnant leaves are simple, ovate to lanceolate, pinnately straight-veined; coarsely toothed.

* **K. japonica** (L.) DC. Japanese rose. Disturbed sites. Commonly cultivated, native of Japan, and occasionally escaping across KY and TN. Plates 260 and 261.

**Koelreuteria** Laxm. Family Sapindaceae. (for Joseph Gottlieb Kölreuter, 1733–1806, Professor of Natural History, Karlsruhe, pioneer hybridizer). Tree to 12 m tall; bark gray-scaly; twigs moderately stout, zigzag, with conspicuous orange lenticels, these encircled by a pale ring; terminal buds absent; buds conical, 2-scaled; leaf scars alternate, shield-shaped, bundle scars > 3 and clustered in center of scar; stipular scars absent; fruit a bladdery, thin-walled capsule, about 5 cm long, splitting open to reveal 3 small black seeds.

** **K. paniculata** Laxm. Golden rain-tree. Commonly cultivated from eastern Asia, and occasionally escaping, at least in IP of KY and and IP and VR of TN. An Alert species in TN. (*paniculatus, panilosus* = with a branched-racemose or cymose inflorescence, tufted, paniculate). Plates 262 and 263.

**Leiophyllum** R.Hedw. Family Ericaceae. (*leiophyllus* = smooth leaf). Low evergreen shrub to 2 m tall, the bark shreddy; twigs lined with decurrent petiole bases, leaf scars with single bundle scar; leaves mostly alternate, sometimes opposite, simple, elliptic, < 20 mm long, entire; stipular scars absent; fruit a dry capsule, 3–4 mm long.

**L. buxifolium** (P.J.Berg.) Elliott. Sand-myrtle. Rocky shrub/heath balds. BR of TN. FACU. (*buxus-folium* = box leaved). [*Kalmia buxifolia* (P.J.Berg) Gift, Kron, & Stevens]. Plate 264.

**Lespedeza** Michx. Family Fabaceae. (for V.M. de Céspedez, eighteenth-century Spanish Governor of Florida). Shrub to 3 m tall, often multi-stemmed, the twigs white-lined; terminal buds absent; lateral buds small and ovoid, reddish and scaly, superposed and collateral; leaf scars alternate, raised, the bundle scars obscure; stipules deciduous; fruit a 1-seeded legume.

** **L. bicolor** Turcz. Bush clover, shrub lespedeza. Infrequent, native of Japan, cultivated for wildlife food, and occasionally escaped and naturalized across KY and TN. Significant Threat in KY and Severe Threat in TN. *L. thunbergii* (DC) Nakai, differing in more drooping fruiting clusters and calyx lobes longer than the tube, is listed as a Lesser Threat in KY. (*bicolor* = of two colors). Plates 265 and 266.

☠ **Leucothoe** D.Don. Dog-hobble. Family Ericaceae. (an ancient Greek name, *Leucothoe* was daughter of King Orchanus of Babylon, and loved by Apollo). Evergreen or deciduous shrubs; twigs with flower buds in slender racemes, these present in winter; terminal buds absent; leaf scars alternate, half-round, with single bundle scar; stipular scars absent; evergreen species with leaves simple, leathery, ovate-lanceolate, serrate and acuminate; fruit a 5-parted capsule with persistent style, in terminal or axillary racemes. All parts are likely toxic.

1. Leaves evergreen, serrate_____**L. fontanesiana.**
1. Leaves deciduous (remnant leaves with low teeth).
    2. Fruiting axis straight, the capsules rounded between the sutures; flowering racemes with leafy bracts _____**L. racemosa.**
    2. Fruiting axis curved, the capsules indented between the sutures; flowering racemes lacking leafy bracts _____**L. recurva.**

**L. fontanesiana** (Steud.) Sleumer. Highland d.-h. Mixed mesophytic forests, often along streams. AP and BR of TN. (*fontanesiana* = of fountains or fast-moving streams). Plates 267, 268, and 269.

! **L. racemosa** (L.) A.Gray. Swamp d.-h., fetterbush. Wet woods and stream banks. Known only from 3 counties in C and E TN. Threatened in TN. FACW. (*racemosa* = raceme-like). [*Eubotrys racemosa* (L.) Nutt.]. Plate 270.

! **L. recurva** (Buckley) A.Gray. Redtwig d.-h. High-elevation forests. S AP of KY and BR of TN. Endangered in KY. FACU. (*recurvo, recurvere* = curved backwards, recurved). [*Eubotrys recurva* (Buckley) Britton]. Plates 271 and 272.

☠ **Ligustrum** L. Privet. Family Oleaceae. (*Ligustrum* = ancient European name for privets). Deciduous or semi-evergreen shrubs; twigs often pubescent; terminal buds present, small with sharp-tipped scales; leaf scars or leaves opposite, the leaves simple, ovate to elliptic, entire, the leaf scars elliptic, not connected by lines, with single bundle scar; stipular scars absent; fruit a blackish drupe with calyx remnant at base of fruit, in terminal panicles. Fruits are attractive, but should not be eaten—ingestion can lead to gastrointestinal problems, convulsions, and death.

1. Twigs densely hairy; leaves hairy at least on the midrib below _____**L. sinense/L. obtusifolium**
1. Twigs slightly hairy; leaves glabrous _____**L. vulgare.**

\*\* **L. obtusifolium** Siebold & Zucc. Border p. This species is indistinguishable from *L. sinense* in winter. It is now less frequent, but is becoming increasingly invasive in recent years in C KY and TN (on the Alert list in TN), from Japan. (*obtusifolium* = blunt-leaved). Not pictured.

\*\* **L. sinense** Lour. Chinese p. Disturbed sites and woodlands. Across KY and TN, from China. This species has become widely naturalized in the region. Severe Threat in KY and TN. (*sinense* = from China). Plates 273, 274, and 275.

\*\* **L. vulgare** L. European p. Disturbed sites and woodlands. Across KY, from Europe. Severe Threat in KY and TN. (*vulgare* = common). Not pictured.

Other privet species occasionally escape in KY and TN, including *L. ovalifolium* Hassk, *L. amurense* Carrière, both being semi-evergreen, and *L. japonicum* Thunb., which has glossy evergreen leaves. The latter species is currently on the Alert list in TN.

🍃 **Lindera** Thunb. Family Lauraceae. (for Johann Linder, 1678–1723, Swedish botanist). Shrub to 4 m tall, with spicy-aromatic fragrance in twigs and bark; twigs greenish, with large pith; end buds may appear as either true or false; buds scaly, about 2 mm long, often superposed or collateral, the leaf buds ovoid and the flower buds globose and stalked; leaf scars alternate, half-round, with 3 bundle scars (or obscured); stipular scars absent; fruit a red drupe to 10 mm long, with calyx remnant at base of fruit. Remnant leaves are simple, elliptic to obovate, pinnately veined, and entire. Twigs, bark, and fruit are edible (dried fruits can be used as a substitute for allspice and a tea can be prepared from the twigs and bark).

**L. benzoin** (L.) Blume. Spicebush. Mesic woods and stream banks. Across KY and TN. Frequent. FACW–, FACW. (*benzoin-* = from an Arabic or Semitic name, *luban-jawi*, signifying Javanese perfume or gum). Plates 276 and 277.

**Linnaea** L. Family Caprifoliaceae or Linnaeaceae. (for Carolus Linnaeus, 1707–1778, author of *Species Plantarum*, published in 1753, initiating era of binomial nomenclature). Trailing, barely woody evergreen, with few erect branches to 10 cm tall, these terminated by a fruiting peduncle to 10 cm long; leaves oval to rounded, 1–2 cm long, opposite and few-toothed; fruit dry, indehiscent, and 1-seeded.

! **L. borealis** L. Twinflower. Mountain woods of E TN, known only from one century-old historical record. Endangered, possibly extirpated in TN. Plate 278.

☤ **Liquidambar** L. Family Hamamelidaceae or Altingiaceae. (*Liquidambar* = liquid-amber, referring to the aromatic resin in the bark). Tree to 35 m tall, with resinous sap exuding from cut twig, the bark rough, thickly ridged and furrowed; twigs glossy, slightly aromatic, often with corky wings, especially on younger trees, and stubby spur branches often present; terminal buds present, imbricate scaly, to 12 mm long, glossy brown and green; leaf scars alternate, elliptic to triangular, with 3 bundle scars, each scar pale with a dark center; stipular scars absent; fruit mace-like, a ball-shaped multiple of beaked capsules, to 4 cm broad. Remnant leaves are simple, star-shaped, and toothed. Soaked twigs were used by Native Americans to clean the teeth. The boiled leaves and bark were used for treating skin inflammation and wounds. The bark exudate has been used as a substitute for chewing gum.

**L. styraciflua** L. Sweetgum. Bottomland to upland woods. Across KY and TN. Frequent. FAC, FAC+. (*styracifluus* = flowing with gum). Plates 279 and 280.

**Liriodendron** L. Family Magnoliaceae. (*Liriodendron* = lily-tree). Tree to 50 m tall, the bark grayish-brown and deeply furrowed in older trunks; twigs slightly aromatic; pith diaphragmed; terminal buds present, with valvate scales, the terminal bud much larger than laterals, with 2 valvate scales, duckbill-like, about 1 cm long; leaf scars alternate, elliptic, with many bundle scars; stipular scars encircling the twig; fruit a samara in cone-like aggregates. Remnant leaves are simple, pinnately veined, 4-lobed. This species reaches the greatest height, and is one of the fastest growing, of any deciduous tree in eastern U.S.

**L. tulipifera** L. Tuliptree, tulip-poplar, yellow-poplar. Rich woods. Across KY and TN. Frequent. FACU, FAC. (*tulipiferus* = having tuliplike flowers). Plates 281, 282, and 283.

☠ **Lonicera** L. Honeysuckle. Family Caprifoliaceae. (for Adam Lonitzer, 1528–1586, German physician and botanist). Woody vines or shrubs; twigs with pith continuous or hollow or excavated; buds scaly, the scales in opposite pairs, varying from short- to long-acuminate, the bud scales persistent at twig bases; terminal buds present, scaly, lateral buds sometimes superposed; leaf scars opposite, connected by lines, with 3 bundle scars; stipular scars absent; fruit a red or black berry with calyx remnant at apex of fruit. Remnant leaves are simple, ovate to oblong, pinnately veined, and entire. The berries are not edible, although attractive, and may cause vomiting, diarrhea, convulsions, or death.

1. Plant a woody vine; berries black or red.
    2. Branches pubescent; leaves membranous but often persistent into winter, the upper leaves not connate-perfoliate, but sometimes pinnately lobed; berries black _____ **L. japonica.**
    2. Branches glabrous, the upper leaves connate-perfoliate; berries red.
        3. Leaves often present in winter; fruiting stalk to 5 cm, the flower and fruit clusters spikelike; leaves green above and glaucous below _____ **L. sempervirens.**
        3. Leaves deciduous (but remnant leaves occasionally present); fruiting stalk to 2 cm, the fruit clusters headlike; remnant leaf surfaces variable (see species accounts below) _____ **L. dioica/L. flava/L. reticulata.**

1. Plant an erect, many-stemmed shrub; berries red.
    4. Twigs glabrous with continuous pith.
        5. Leaves deciduous; rare native shrub _____ **L. canadensis.**
        5. Leaves persistent in winter, with a bristle-tip; non-native species _____ **L. fragrantissima.**
    4. Twigs pubescent and hollow.
        6. Fruit stalks shorter than petioles _____ **L. maackii.**
        6. Fruit stalks longer than the petioles _____ **L. morrowii.**

! **L. canadensis** Bartr. Fly h. High-elevation woodlands. BR of TN. FACU. Rare. Threatened in TN. Not pictured.

! **L. dioica** L. Wild h. Mesic woodlands, in thickets or on rocky ledges. In C and E KY (very rare) and TN (infrequent). FACU. Three varieties based on leaf features: var. **dioica**, var. **glaucescens** (Rydb.) Butters, and var. **orientalis** Gleason. The latter variety is currently listed as Endangered in KY and the species is listed as Special Concern in TN. (*dioicus* = of two houses). Remnant leaves in this species are green above and glaucous below. Plate 284.

! **L. flava** Sims. Yellow h. Rocky woodlands and thickets of S TN (1 county). Threatened in TN. (*flava* = yellow). Remnant leaves in this species are green on both surfaces. Plate 285.

** **L. fragrantissima** Lindl. & Paxton. Sweet-breath-of-spring. Disturbed places, in cultivation. Escaped at few sites across KY and TN. An Alert List species in TN. Introduced from China. (*fragantissimus* = most fragrant). Plates 286 and 287.

** **L. japonica** Thunb. Japanese h. Thickets, fencerows, disturbed places. Across KY and TN. Abundant, naturalized from Asia. Severe Threat in KY and TN. FAC–. Plates 288, 289, 290, and 291.

** **L. maackii** (Rupr.) Maxim. Amur h. Disturbed places, fencerows, woodlands. Chiefly in C KY and TN, scattered across both states. Naturalized from Asia, and often abundant in the understory. This species is one of our worst woody weeds, especially in limestone areas. Severe Threat in KY and Significant Threat in TN. (*maackii* = for Richard Maack, 1825–1886, Russian naturalist). Plates 292, 293, and 294.

** **L. morrowii** A.Gray. Morrow's h. Disturbed places. Escaped at few scattered sites in central and E KY and TN. Rare, naturalized from Japan. Severe Threat in KY and Alert List in TN. (*morrowii* = named by Asa Gray for Dr. James Morrow, who discovered the plant on a trip to Japan in early 1860s). Not pictured.

! **L. reticulata** Raf. Grape h. Mesic wooded slopes, often over ledges. IP. Known only from 4 counties KY and 2 counties in TN. Threatened in KY and Endangered in TN (*reticulata* = veiny). Remnant leaves are glaucous on both surfaces. [*L. prolifera* (G.Kirchn.) Rehder]. Not pictured.

**L. sempervirens** L. Trumpet h. Mesic woods and thickets, frequently cultivated. Across KY and TN. Infrequent. FACU, FAC. (*sempervirens* = always green). Plate 295.

☠ **Lycium** L. Family Solanaceae. (a thorn tree from Lycia, Turkey). Scrambling vines or shrubs to about 3 m tall; twigs grayish, lined, often bearing spines; buds small, subglobose, often clustered; leaf scars alternate, raised, with single bundle scar; stipular scars absent; fruit a poisonous red berry with calyx remnant at base of fruit.

* **L. barbarum** L. Matrimony-vine. Disturbed places. Across KY and TN. Infrequent, cultivated and escaping, native of Eurasia. (*barbarus, barbaria* = foreign, from Barbary). Plate 296.

☠ **Lyonia** Nutt. Family Ericaceae. (for John Lyon, ca. 1765–1814, introducer of American plants). Shrub to 3 m tall; terminal buds absent; lateral buds ovoid and long-pointed, about 5 mm long, 2-scaled; leaf scars alternate, half-round, with single bundle scar; stipular scars absent; fruit a rounded, 5-sectioned capsule, about 4 mm wide, persistent. Remnant leaves are ovate to obovate, pinnately veined, entire to slightly toothed. All parts considered toxic.

**L. ligustrina** (L.) DC. Maleberry. Swamp forests and stream banks. E TN and S AP of KY. Infrequent. FACW. Two varieties: var. **foliosiflora** (Michx.) Fernald, with the umbel-like inflorescence conspicuously bracteate; and var. **ligustrina**, with few or no bracts on the inflorescence, are recognized by some authors. (*ligustrina* = privetlike, resembling *Ligustrum*). Plates 297 and 298.

☠/➤ **Maclura** Nutt. Family Moraceae. (for William Maclure, 1763–1840, American geologist). Trees with rounded crowns, outer bark thickly ridged, inner bark yellow, and wood orangish, as are the exposed roots; twigs often with thorns above the leaf scars, spur branches, and milky juice; terminal buds absent; lateral buds scaly, often paired, rounded and glabrous, partly concealed at thorn bases; leaf scars alternate, half-round or nearly round, with > 3 bundle scars; stipular scars present but small; fruit a multiple of achenes, similar in size to a grapefruit. Remnant leaves are simple, ovate, pinnately veined, long-petioled, tapered apex, with entire margin. Osage-orange has a number of uses, and there are several websites maintained by devotees of this tree detailing such uses as how the trees can be used as natural fences, bark can be used for a dye and as a source of fiber, the wood as fuel, the branches for making excellent bows, the fruits for repelling pests, and many more! The milky sap is considered toxic.

\*\* **M. pomifera** (Raf.) C.K.Schneid. Osage-orange, hedge-apple, horse-apple, bodark, bodock. Fencerows and disturbed places. Across KY and TN. Frequent, especially in limestone areas, introduced from E Texas and E Oklahoma region primarily for fencerows, now widely naturalized in E U.S. Significant Threat in TN. UPL, FACU. (*pomifer* = pome-bearing, bearing applelike fruits). Plates 299, 300, and 301.

**Magnolia** L. Magnolia. Family Magnoliaceae. (for Pierre Magnol, 1638–1715, Professor of Botany and Director of Montpellier Botanic Garden). Small to large trees, evergreen in some; twigs with diaphragmed pith; terminal bud present, single-scaled (2 fused stipules); leaf scars alternate, rounded or U-shaped, with many bundle scars; stipular scars encircling the twig at nodes; fruit a cone-like aggregate of follicles with red-arillate seeds.

1. Leaves evergreen, oblong to elliptic to oblanceolate, either reddish or silvery below, tapered at both ends, mostly < 20 cm long.
    2. Twigs and lower leaf surfaces rusty-pubescent _____ **M. grandiflora.**
    2. Twigs and lower leaf surfaces smooth or with pale pubescence _____ **M. virginiana.**
1. Leaves deciduous (remnant leaves otherwise, some with auriculate bases, often much longer).
    3. Terminal buds < 1.8 cm long, silvery silky; trees to 25 m tall, with furrowed bark; leaf scars distributed regularly along twigs; remnant leaves tapered to base _____ **M. acuminata.**
    3. Terminal buds > 2 cm long, pubescence various; understory trees to 15 m tall, with smooth bark; leaf scars clustered toward twig tips; remnant leaves various.
        4. Terminal buds and fruits pubescent; fruits ovoid to globose; remnant leaves with auricles at base _____ **M. macrophylla.**
        4. Terminal buds and fruits glabrous; fruits elliptic to cylindric; remnant leaves various.
            5. Remnant leaves tapered to base _____ **M. tripetala.**
            5. Remnant leaves with auricles at base _____ **M. fraseri.**

**M. acuminata** (L.) L. Cucumber m. Mixed mesophytic forests. In E TN and in AP of KY, also in far-western counties of KY and TN on loess bluffs above Mississippi River. Infrequent. (*acuminata* = with a long, narrow and pointed tip). Plate 302.

**M. fraseri** Walter. Fraser m. Mixed mesophytic forest. Chiefly BR of TN and SE AP of KY. Infrequent. FACU, FAC. (*fraseri* = for John Fraser, 1750–1811, nursery man of Chelsea, England). There is an historical record of *Magnolia pyramidata* Bartr. from W KY; it differs from *M. fraseri* in its smaller fruits, < 6 cm long, and is more associated with the Coastal Plain of SE U.S. Plates 303 and 304.

* **M. grandiflora** L. Southern m. Native to S U.S., and rarely escaping across KY and TN. FACU, FAC+. (*grandiflora* = with large flowers). Plate 305.

**M. macrophylla** Michx. Bigleaf m. Mixed mesophytic forest. AP, BR, and IP of TN; AP and IP (Edmonson County only) of KY. Frequent. (*macrophylla* = with large leaves). Plates 306, 307, and 308.

**M. tripetala** L. Umbrella m. Mixed mesophytic forest. Through eastern half of TN, also some W IP counties; in AP of KY and few IP counties. Frequent. FACU, FAC. (*tripetalus* = three petals). Plates 309 and 310.

! **M. virginiana** L. Sweetbay. Forested wetlands. S TN. Threatened in TN. FACW+. Plate 311.

**Mahonia** Nutt. Family Berberidaceae. (after Bernard McMahon, 1775–1816, American horticulturist). Evergreen shrub, to 2 m tall, the inner bark yellowish; leaves pinnately compound, the leaflets leathery and spiny-margined, sessile on the rachis; fruit a blackish berry in racemes.

** **M. bealei** (Fortune) Carrierre. Oregon grape-holly. Disturbed sites. An Alert List species in TN. Occasional escape, a native of China. (*Berberis bealei* Fortune). (*bealei* = for Thomas C. Beale, nineteenth-century supporter of Robert Fortune's botanical work). Plate 312.

**Malus** Mill. Apple, crabapple. Family Rosaceae. (*Malus* = the ancient Latin name for an apple tree). Shrubs or small trees, sometimes thorny, and bark often gray-scaly; twigs bearing spur-branches, these sometime thorn-like; terminal buds present, with 4 or 5 scales, the lateral buds similar but much smaller; leaf scars alternate, linear, with 3 bundle scars; stipular scars lacking; fruit a pome, but absent in winter. The fruits are edible and nutritious, but the seeds contain a cyanide-like compound, and should not be ingested in quantity. The sturdy branches are useful for making bows.

1. Buds completely gray-hairy; thorns absent; fruit > 5 cm broad; remnant leaves serrate and unlobed _____**M. pumila.**
1. Buds hairy but mostly at base on along scale margins; thorns or thorny branches present; fruit < 5 cm broad; remnant leaves serrate and often lobed.
    2. Remnant leaves pubescent below _____**M. ioensis.**
    2. Remnant leaves glabrous below.
        3. Leaves crenate-serrate, obtuse _____ **M angustifolia.**
        3. Leaves doubly serrate, acute to acuminate _____ **M. coronaria.**

**M. angustifolia** (Aiton) Michx. Southern c. Dry to mesic woods and thickets, upland and lowland. Across KY and TN, but less common in C KY and TN. (*angustifolius* = narrow leaved). Plate 313.

**M. coronaria** (L.) Mill. Sweet c. Dry to mesic woods and thickets, uplands and lowlands. IP and AP of KY and TN, nearly absent from central parts of states. Infrequent. (*corona* = crown material). Plate 314.

! **M. ioensis** (Alph.Wood) Britton. Iowa c. Oak barrens. CP and W IP of KY. Rare. Special Concern in KY. (*ioensis* = from Iowa). Not pictured.

\* **M. pumila** Mill. Common a. Cultivar from Eurasia, sporadically escaping to disturbed sites, or long persistent, across C and E KY and TN. (*pumilio* = very small, low, small, dwarf). Plate 315.

☠ **Melia** L. Family Meliaceae. (*Melia* = for the Greek name for ash tree). Tree to 15 m tall, with bark scaly and ridged; twigs stout, with pale lenticels; terminal bud absent; lateral buds rounded and hairy, appearing naked; leaf scars alternate, 3-lobed, with 3 U-shaped bundle scars (or 3 clusters of scars); stipular scars absent; fruit a yellow drupe. The fruits are tempting, colorful, fleshy, and abundant, and some people have succumbed to the temptation, but ingestion has led to nausea, paralysis, and death! All parts considered highly poisonous.

\*\* **M. azedarach** L. Chinaberry. A native of Asia, cultivated and escaping in S U.S., documented from several counties in W TN, and from a single county in E TN, near the KY border. Significant Threat in TN. (*azedarach* = a Middle Eastern vernacular name for the bead tree). Plates 316 and 317.

☠ **Menispermum** L. Moonseed. Family Menispermaceae. (*Menispermum* = moonseed). Woody twining vine; twigs striate, the buds partly buried in stem, hairy, clustered at the nodes; leaf scars alternate, circular or notched, with concave center, the bundle scars 3–7; stipular scars absent; fruit a blackish drupe to 8 mm long, with single semicircular seed. Remnant leaves are simple, orbicular, palmately veined and usually lobed, the petiole apex attached to leaf base slightly inside the leaf margin. All parts highly toxic and the ingestion of fruits may have fatal consequences. Note that the fruits superficially resemble those of edible wild grapes, but unlike moonseed fruits, wild grapes are many-seeded!

**M. canadense** L. Common m. Rich woods and thickets. Across KY and TN. Frequent. Plates 318 and 319.

☠ **Menziesia** Smith. Family Ericaceae. (for Archibald Menzies, 1754–1842, Scottish naturalist who sailed on the HMS *Discovery* with Captain George Vancouver, 1791–1795). Shrub to 2.5 m tall, with shreddy bark; twigs finely hairy and with longer bristly hairs; terminal buds present, about 7 mm long, sometimes clustered at twig tips; buds with about 6 imbricate scales, the margins white-ciliate; leaf scars alternate, with single bundle scar; stipular scars absent; fruit an ovoid capsule to 7 mm long, persistent. Remnant leaves are simple, elliptic to obovate, pinnately veined, with jagged scales on midvein of lower leaf surfaces. All parts are likely toxic.

! **Menziesia pilosa** (Michx.) Juss. Minnie-bush. Thin woods and balds. BR of TN. Infrequent. FACU. Special Concern in TN. (*pilosa* = covered by soft hairs). Plates 320 and 321.

✿ **Mitchella** L. Family Rubiaceae. (for Dr. John Mitchell, 1676–1768, botanist in Virginia). Creeping subshrub; leaves evergreen, opposite, to 2 cm long and wide, ovate; fruit red and berry-like, edible and persistent.

**M. repens** L. Partridge-berry. Mesic woods. Across KY and TN. Frequent. FACU, FACU+. (*repens* = creeping). Plate 322.

✿/☠/➤ **Morus** L. Mulberry. Family Moraceae. (*morus* = the ancient Latin name for the mulberry). Trees to 20 m tall, but usually < 10 m tall, with broad crowns and scaly or ridged barked; twigs with milky sap; terminal buds absent; lateral buds to 7 mm long, with 5–7 imbricate scales; leaf scars alternate, nearly circular, in one plane, the leaf scars depressed in center, with numerous bundle scars, sometimes in a ring pattern; stipular scars linear, unequal; fruit a fleshy multiple, blackberry-like, but absent in winter. Remnant leaves are simple, ovate to orbicular, palmately veined, sometimes lobed, and toothed. The mature fruits of red mulberry (purple-black when ripe) are edible and very sweet, but the fruits of white mulberry (whitish-pink when ripe) are insipid and should be dried first. Note that the twigs and unripe fruits have a milky sap and these should not be ingested; they may cause stomach problems or other bad reactions, including hallucinations. The bark of white mulberry has been used as a source of fiber for making ropes.

1. Mature bark tight to furrowed, with orangish streaks; lateral buds about as wide as long, blunt and appressed against the twig, the scales with brownish margins _____ **M. alba.**
1. Mature bark scaly, lacking orangish streaks; lateral buds longer than wide, acute and spreading, the scales with dark margins_____ **M. rubra.**

\*\* **M. alba** L. White m. Fencerows, roadsides, other waste places. Across KY and TN. Frequent. Naturalized from Asia. Significant Threat in KY. UPL, FACU–. Plate 323.

**M. rubra** L. Red m. Mesic woods and thickets. Across KY and TN. Frequent. FACU, FAC. (*M. murrayana* D.E.Saar & S.J.Galla). Plate 324.

**Nandina** Thunb. Family Berberidaceae. (*Nandina* = Latinized version of Japanese name for the plant, *nan-ten*). Evergreen shrub to 2 m tall; stems few branched, with leaves at apex of plant; leaves evergreen, bipinnately or tripinnately compound; fruit a red berry, often in large terminal clusters.

\*\* **N. domestica** Thunb. Heavenly-bamboo. Native of Asia, and now documented from several counties in C and E TN; expected to become more invasive; an Alert species in TN. Plates 325 and 326.

**Nemopanthus** Raf. Family Aquifoliaceae. (*Nemopanthus* = thread flower). Shrubs, the twigs typically bearing spur branches; terminal buds present, buds rarely superposed (as in related genus *Ilex*), bud scales 4–6, blunt; leaf scars alternate, crescent-shaped, with single bundle scar; twigs with tiny black stipules or tiny stipular scars; fruit fleshy, berry-like, and red, often persisting into winter. Remnant leaves are similar to those of deciduous hollies.

! **N. collinus** (Alexander) R.C.Clark. Appalachian mountain holly. Known only from high elevations of Sevier County in BR of TN (also in few counties in WV, VA, and NC). (*collinus* = of the hills). Special Concern in TN. (*Ilex collina* Alexander). Plate 327.

**Nestronia** Raf. Family Santalaceae. (*Nestronia* = from the Greek name of *Daphne*). Shrubs to 1.5 m tall, parasitic on tree roots; twigs expanded at the nodes, pubescent, the pith brown; terminal buds absent; buds reddish, with keeled and imbricate scales; leaf scars opposite, not connected by lines, half-round, with single bundle scar; fruit a greenish drupe to 1 cm long. Remnant leaves are simple, lance-ovate, pinnately veined, and entire.

! **N. umbellula** Raf. Conjurer's-nut. Dry upland woods. S IP of KY, E IP and BR of TN. Endangered in KY and TN. (*umbellulatus* = umbel-like). Plates 328 and 329.

**Neviusia** A.Gray. Family Rosaceae. (named by Asa Gray for its discoverer, Reverend R.D. Nevius, a missionary of the Protestant Episcopal Church, who collected the plant in 1857 in northern Alabama). Shrub to 2 m tall, stoloniferous; twigs slender, pubescent with white trichomes, the twigs often dying back in winter; terminal buds absent; buds with imbricate scales; bundle scars 3; stipules dark, leafy and often persistent; fruit a fleshy achene. Remnant leaves are simple, ovate to lanceolate, pinnately veined, and sharply toothed.

! **N. alabamensis** A.Gray. Alabama snow-wreath. Known from few sites in C TN, in bouldery areas or above streams, on limestone or shale soils. Threatened in TN. Plates 330 and 331.

**Nyssa** L. Family Nyssaceae. (*Nyssa* = water nymph). Trees to 40 m tall, with ridged to blocky bark; twigs with short spur branches and diaphragmed pith; terminal buds present, scaly imbricate, to 9 mm long, often with purple and green coloration, the lateral buds smaller and often superposed; leaf scars alternate, crescent to triangular, with 3 bundle scars; stipular scars absent; fruit a blackish drupe, falling in autumn. Remnant leaves are simple, elliptic to obovate, pinnately veined, and usually entire, but sometimes with a few teeth.

1. Trees with buttresses, standing in or near water; buds and drupes various.
    2. Buds obtuse, to 4 mm long; drupes 2–4 cm long, the stones sharply winged _____ **N. aquatica.**
    2. Buds acute, to 7 mm long; drupes 1–1.5 cm long, the stones faintly ribbed _____ **N. biflora.**
1. Trees lacking buttresses, the habitat varying from lowland to upland, but usually not in standing water; buds acute, to 9 mm long; drupes 1–1.5 cm long, the stone faintly ribbed _____ **N. sylvatica.**

**N. aquatica** L. Water tupelo. Swamp forests. CP of KY and TN, and few sites in south-central TN. Infrequent. OBL. (*aquaticus* = living in water). Plate 332.

**N. biflora** Walter. Swamp tupelo. Swamps. Rare in CP of KY, more common in CP of TN, with few scattered sites in C TN and KY. FACW+, OBL. (*biflorus* = two-flowered). Plate 333.

**N. sylvatica** Marshall. Blackgum. Wet to dry forests. Across KY and TN. Frequent. FAC. (*sylvaticus* = wild, of woods or forest). Plates 334, 335, and 336.

**Opuntia** Miller. Prickly-pear. Family Cactaceae. (*Opuntia* = Tournefort's name for succulent plants). Stem a flattened, fleshy, green pad, with clusters of needlelike spines and tiny barbed bristles; fruit a purplish berry to 5 cm long. Fruits and pads of prickly-pear are edible, being careful to avoid tiny spines. The fruits mature in the fall and the inner flesh can be scooped out; pads can be peeled or sliced and roasted or fried. The sliced pads can be used as a compress for wounds or as a source of water.

**O. humifusa** (Raf.) Raf. Eastern p.-p. Rocky limestone glades, bluffs, and open woodlands. Chiefly IP in KY, across TN, but more frequent in C TN, especially on river bluffs. [*O. compressa* J.F.Macbr.]. (*humifusus* = trailing, sprawling, spreading over the ground). Plate 337.

**Ostrya** Scop. Family Betulaceae. (*Ostrya* = hard scale). Small tree to 20 m tall, with thinly shreddy bark; twigs slender and zigzag, bearing staminate catkins; terminal buds absent; buds scaly-imbricate to 6 mm long, narrowly ovoid, the scales striate, > 6, usually light green with dark margins; leaf scars alternate, in one plane, crescent to elliptic, with 3 bundle scars; stipular scars narrow, unequal; fruit a nutlet subtended by 3-lobed leafy bracts. Remnant leaves are simple, ovate to elliptic, pinnately veined, and double-toothed.

**O. virginiana** (Mill.) K.Koch. Hophornbeam, ironwood. Dry to mesic forest. Across KY and TN. Frequent. FACU–. Plates 338 and 339.

**Oxydendrum** DC. Family Ericaceae. (*Oxydendrum* = sour tree). Tree to 25 m tall, but usually much shorter, with deeply furrowed to blocky bark; twigs slender, zigzag, often reddish or greenish; terminal buds absent; buds scaly and rounded, to 3 mm long, partly sunken; leaf scars alternate, shield-shaped, with single bundle scar; stipular scars absent; fruit a small 5-valved capsule in terminal racemes, present in winter. Remnant leaves are simple, ovate to elliptic, pinnately veined, and finely toothed to nearly entire. Sourwood honey, with a syrupy and spicy flavor, is one of the best tasting and most prized of all honeys. Pure sourwood honey is hard to obtain, as it is difficult to prevent the the bees from gathering nectar from other species with overlapping flowering periods (e.g., sumacs). In addition, it is becoming increasingly difficult to find good stands of sourwood trees, and the trees are highly sensitive to environmental conditions, requiring adequate sunlight and rainfall to produce high-quality nectar.

**O. arboreum** (L.) DC. Sourwood. Dry to mesic forest, from ridges to swamps. Across KY and TN, but much less frequent in the CP. Frequent. (*arboreus* = tree like, branched). Plates 340 and 341.

⚘ **Parthenocissus** Planch. Family Vitaceae. (*Parthenocissus* = virgin-ivy). Woody vines climbing by branched tendrils, these attached opposite the leaf scars and tipped with adhesive discs; twigs with conspicuous lenticels; terminal buds absent; lateral buds small, conical, scaly; leaf scars alternate, raised and circular, with numerous but indistinct bundle scars; stipular scars narrow and elongate; fruit a bluish-black berry, maturing in late summer. Virginia-creeper is common and the berries may appear tempting, but ingestion is likely to cause digestive problems and drowsiness, and possibly other more serious consequences (may be fatal in children).

**P. quinquefolia** (L.) Planch. Virginia-creeper. Mesic to wet woods and openings, various habitats. Across KY and TN. Frequent. FACU, FAC. (*quinquefolia* = five leaved). Plate 342.

**Paulownia** Siebert & Zucc. Family Scrophulariaceae or Bignoniaceae, or more recently placed in the Paulowniaceae. (for Princess Anna Paulovna, 1795–1865, consort of King William II of the Netherlands, and daughter of Czar Paul I of Russia). Tree to 15 m tall, with thin, flaky bark; twigs stout, the pith large and chambered in the internodes; terminal buds absent; lateral buds hemispheric, partly sunken, superposed, 4-scaled; leaf scars opposite, not connected by lines, circular, with numerous bundle scars in an ellipse; fruit an ovoid capsule to 4 cm long, in terminal panicles. Remnant leaves are simple, cordate, to 30 cm or more long, softly pubescent on lower surface, palmately veined, and entire or slightly lobed. Stalks of brownish-hairy flower buds often present in winter. The princess-tree is highly invasive but also prized for its rapid growth and easily-worked wood. The species is also used for strip-mine reclamation efforts in the U.S.

** **P. tomentosa** (Thunb.) Steud. Princess-tree. Disturbed places and open woodlands. Infrequent across both KY and TN. UPL, FACU. Naturalized from China. Significant Threat in KY and Severe Threat in TN. (*tomentosus* = thickly matted with hair). Plates 343, 344, 345, and 346.

**Paxistima** Raf. Family Celastraceae. (*Paxistima* = thick stigma). Evergreen shrub, stoloniferous, with aerial branches rising to 40 cm high; twigs 4-sided; leaves opposite, simple, < 3 cm long, linear-oblong and serrate; leaf scars with single bundle scar, and stipular scars absent; fruit a capsule to 4 mm long.

! **P. canbyi** A.Gray. Canby's mountain-lover. Dry woodland ridges and limestone cliff lines. Known only from 6 counties in KY and 1 county in TN, chiefly AP. Threatened in KY and Endangered in TN. (*canbyi* = for William Marriott Canby, 1831–1904, American botanist). Plate 347.

**Philadelphus** L. Mock-orange. Family Hydrangeaceae. (*Philadelphus* = brotherly love). Shrub to 3 m tall; twigs lined or angled, often with loose bark; buds with 2 or 4 scales, end buds various; leaf scars opposite, connected by a line or ridge, half-round to triangular, with 3 bundle scars; stipular scars absent; fruit an obovoid capsule to 9 mm broad, splitting into sections. Remnant leaves are simple, lanceolate to ovate, somewhat palmately veined, with margins varying from nearly entire to sharply toothed.

1. Terminal buds usually present; axillary buds evident throughout winter; bark peeling _____ **P. hirsutus.**
1. Terminal buds usually absent; axillary buds hidden under leaf scar (green buds bursting through leaf scar late in winter); bark varying as described below.
   2. Bark gray, not peeling _____ **P. pubescens.**
   2. Bark brown or gray and peeling _____ **P. inodorus.**

**P. hirsutus** Nutt. Cumberland m.-o. Along creeks, banks, and bluffs, especially limestone areas. S IP and AP of KY, across C and E TN. Infrequent. (*hirsutellus* = somewhat hairy, with very short hairs). Plate 348.

! **P. inodorus** L. Appalachian m.-o. Along creeks, banks, and bluffs, especially limestone areas. Rare across KY, infrequent across TN. Threatened in KY. (*inodorus* = without smell, odorless). Plates 349 and 350.

! **P. pubescens** Loisel. Ozark m.-o. Along creeks, banks, and bluffs, especially limestone areas. Rare in W and S IP of KY, infrequent across C and E TN. Endangered in KY. (*pubescens* = hairy). Plate 351.

☠ **Phoradendron** Nutt. Mistletoe. Family Viscaceae or Santalaceae. (*Phoradendron* = tree-burden). Evergreen shrub, parasitic on the branches of a variety of hardwood trees; twigs green and jointed; leaves opposite, simple, to 5 cm long, oblong or obovate, entire; fruit a whitish drupe, with sticky flesh, present in winter. The fruits and leaves are very toxic, potentially fatal, and should not be used to make teas or other herbal preparations.

**P. leucarpum** (Raf.) Reveal & M.C.Johnst. subsp. **leucarpum**. Mistletoe. Parasitic on a variety of host trees, especially rough-barked trees such as black cherry, black walnut, hickories, and American elm. The frequency of particular hosts may shift from region to region. Across both KY and TN. (*leucarpum* = white-fruited). Plate 352.

**Physocarpus** Maxim. Family Rosaceae. (*Physocarpus* = bladder fruit). Shrub to 3 m tall, with shreddy bark; twigs slender, zigzag, ridged below nodes; pith large and brownish; terminal bud present, with about 5 overlapping scales, pointed, to 7 mm long; leaf scars alternate, raised, 3-lobed, with 3 bundle scars, the lower one largest; stipular scars small; fruit a follicle, in umbrella-like clusters. Remnant leaves are simple, ovate or rounded, palmately veined, irregularly toothed, and often lobed.

**P. opulifolius** (L.) Maxim. Common ninebark. Limestone cliffs, stream banks. Across C and E KY and TN. Infrequent. FACW–, FAC–. (*opulifolius* = with leaves resembling the guilder rose). Plates 353, 354, and 355.

**Picea** A.Dietr. Spruce. Family Pinaceae. (*Picea* = pitch). Evergreen tree to 35 m tall, the bark reddish and scaly; leaves spirally arranged, needlelike, 4-angled, sharp-pointed, 12–15 mm long, dark green, sessile on woody stubs, the older twigs with a bumpy appearance due to the persistent woody stubs; cones elongate-ovoid, to 4 cm long, the scales thin and rounded, unarmed. The bark is a good source of cordage, and the species has similar food and medicinal uses as those of Fraser fir.

**P. rubens** Sarg. Red s. At high elevations, BR of TN, above 4500 ft (1370 m), and associating with *Abies fraseri* to form the spruce-fir zone above 5500 ft (1680 m). FACU. Plate 356.

☠ **Pieris** D.Don. Fetterbush. Family Ericaceae. (*Pieris* = from Pierides, name for the muses of Greek mythology). Evergreen shrub to 2 m tall; twigs bristly; buds globose; leaf scar with single bundle scar; leaves alternate, simple, lance-oblong, to 8 cm long, the margins serrulate and ciliate, dotted with dark glands; stipular scars absent; fruit a globose capsule to 6 mm wide. All parts are likely toxic.

! **P. floribunda** (Pursh) Benth. & Hook.f. Mountain f. High elevation balds in BR of TN. Threatened in TN. (*floribunda* = many-flowered). Plate 357.

**Pinus** L. Pine. Family Pinaceae. (*Pinus* = the ancient Latin name for a cone-bearing tree). Trees; leaves needlelike, in clusters of 2, 3, or 5, evergreen, enclosed at base by a sheath of scale leaves; cones with woody scales, armed with a prickle in all except *P. strobus*, the seeds winged and borne 2/scale. The seeds can be eaten (but eastern pines have very small seeds), and the needles can be brewed into a vitamin-rich tea. The inner bark can be dried and ground into flour for mixing with other foods, and is also a good source of fiber for ropes and string. The bark resin is useful for healing of wounds and can be used for waterproofing articles or as a glue. To make a glue, cut a small notch in the bark, collect the resin, and heat it. The hardened resin can be used as an emergency tooth filling. Good bows can be constructed from the branches and saplings, but these may break under tension.

1. Needles 5/fascicle; cones unarmed (soft pines) _____ **P. strobus.**
1. Needles 2 or 3/fascicle; cones armed (hard pines).
    2. Needles short, to 9 cm long, and twisted.
        3. Needles 1 mm thick; cones 4–7 cm long, longer than wide, the prickle straight, 1–3 mm long _____ **P. virginiana.**
        3. Needles 2 mm thick; cones 6–9 cm long and wide, the prickle stout and upcurved, hooklike, 3–8 mm long _____ **P. pungens.**
    2. Needles longer, to 23 cm, straight.
        4. Needles 13–25 cm long; cones to 13 cm long, the prickles slender, reflexed and very sharp _____ **P. taeda.**
        4. Needles to 12 cm long; cones to 7 cm long; prickles various.
            5. Needles 1 mm wide; cones longer than wide, thin-scaled, with tiny prickle; twigs glaucous; bark with resin pockets _____ **P. echinata.**
            5. Needles 2 mm wide; cones top-shaped, about as long as wide, thick-scaled, with stout prickle; twigs brown, not glaucous; bark lacking resin pockets _____ **P. rigida.**

**P. echinata** Mill. Shortleaf p. Dry upland woods. Across KY and TN, but less frequent in limestone regions and in far western sections. Frequent. (*echinata* = covered with prickles). Plate 358.

**P. pungens** Lamb. Table Mountain p. Dry mountain slopes and ridges. Chiefly BR of TN. Infrequent. (*pungens* = ending in a sharp point). Plate 359.

**P. rigida** Mill. Pitch p. Dry mountain woods. AP of KY, with few outliers in IP; VR and BR of TN. Infrequent. FACU, FACU–. (*rigidus* = stiff, inflexible). Plate 360.

**P. strobus** L. Eastern white p. Mesic to dry forests. E KY and E TN, with disjunct populations in W IP of both states. Infrequent. FACU. (*strobus* = the ancient name for an incense-bearing tree). Plate 361.

**P. taeda** L. Loblolly p. Sandy lowland forests, fencerows. Across TN and S KY. Native to S TN, less frequent in N TN and S KY, where it has naturalized from plantings. FACU–, FAC. (*taeda* = an ancient name for resinous pine cones used for torches). Plate 362.

**P. virginiana** Mill. Virginia p. Chiefly in dry woods. All provinces of both states except for western-most sections. Frequent. Plate 363.

**Planera** Gmel. Family Ulmaceae. (for J.J. Planer, 1743–1789, Professor of Medicine at Erfurt, Germany). Small tree to 15 m tall, bark platy, often curling up at edges; twigs slender, zigzag, lenticellate; terminal buds absent; lateral buds to 4 mm long, offset to one side of leaf scar, with about 6 visible scales; leaf scars alternate, in one plane, oval, with 3 depressed bundle scars; stipular scars tiny; fruit a brown nutlet maturing in the spring. Remnant leaves are simple, ovate, pinnately straight-veined, rounded to slightly oblique at base, and single-toothed.

**P. aquatica** (Walter) J.F.Gmel. Water-elm, planer-tree. Swamps and damp woods. CP and W IP of KY and TN. Infrequent. OBL. (*aquaticus* = living in water). Plate 364.

**Platanus** L. Sycamore. Family Platanaceae. (*Platanus* = broad crown). Tree to 50 m tall, with outer bark brown, exfoliating to reveal greenish-white inner bark; twigs slender, zigzag; terminal buds absent; lateral buds conical with single cap-like scale, about 8 mm long; leaf scars alternate, narrow and encircling the lateral buds, with bundle scars 5–9; stipular scars encircling the twigs; fruit a ball-like multiple of achenes, each subtended by a circle of brownish hairs, the fruits persistent into winter. Remnant leaves are simple, round in outline, palmately veined and lobed, with irregular teeth, and with hollow petiole base that caps the axillary bud. The sap can be utilized as a source of energy in emergency situations.

**P. occidentalis** L. Sycamore. Wet woods, along streams. Across KY and TN. Frequent. FACW– (*occidentalis* = western, of the West). Plates 365 and 366.

**Polygonella** Michx. Family Polygonaceae. (*Polygonella* = dimunitive of *Polygonum*, meaning many-jointed). Shrub to 1 m tall, with jointed stems and nodal sheaths encircling the stems; leaves linear, about 1 mm wide and thick, to 15 mm long, and sometimes persistent in winter; fruit an achene.

! **P. americana** (Fisch. & C.A.Mey.) Small. Southern jointweed. Sand or gravel bars along streams. Known from Morgan County, TN. Endangered in TN. Plates 367 and 368.

**Poncirus** Raf. Family Rutaceae. (*Poncirus* = the French name for Japanese bitter orange). Shrubs or small tree to 7 m tall, with green bark; twigs green and ridged, bearing thorns with flattened bases; terminal buds absent; lateral buds small and rounded; leaf scars alternate, rounded, with single bundle scar; stipular scars absent; fruit a citrus berry, to 5 cm thick, but not edible (can result in digestive problems, and contact with fruits can cause dermatitis). Leaves, sometimes persistent, are compound with 3 leaflets, and winged petioles.

* **P. trifoliata** (L.) Raf. Trifoliate orange. Disturbed sites. Native of China, cultivated and persistent around home sites, occasionally naturalized across TN. (*trifoliata* = having three leaflets). Plate 369.

**Populus** L. Poplar, cottonwood, aspen. Family Salicaceae. (*Populus* = the ancient name for poplar, tree of the people). Trees, often with root sprouts; twigs rounded to angled, often bearing short spur shoots, and pith angled in cross-section; terminal buds present, imbricate scaly; lateral buds with lowest bud scale about as wide as the twig and centrally located over the leaf scar; leaf scars alternate, raised, crescent to triangular, with 3 bundle scars; stipular scars narrow and distinct; fruit a capsule with white-tufted seeds. Remnant leaves simple, ovate to deltoid, pinnately veined, toothed, with elongate petioles in most (sometimes longer than the blades). The inner bark contains salicin, a precursor of salicylic acid (or aspirin), and an extract was used by Native Americans as a tonic and for healing purposes. The inner bark can be eaten raw in emergency situations, and is also a good source of fiber for cordage.

1. Terminal buds ovoid, < 1.5 cm long; bud surfaces various.
    2. Bark smooth on upper trunk, whitish to greenish; buds not resinous; remnant leaves coarsely toothed, usually sharp-tipped.
        3. Twigs and buds white-tomentose; remnant leaves often lobed, with terete petioles _____ **P. alba.**
        3. Twigs and buds smooth or finely pubescent; remnant leaves ovate and few-toothed, unlobed, the petioles flattened _____ **P. grandidentata.**
    2. Bark furrowed on upper trunk, brownish or grayish; buds resinous; remnant leaves finely toothed, blunt-tipped _____ **P. heterophylla.**
1. Terminal buds lanceolate, > 1.5 cm long, glabrous and sticky-resinous.
    4. Twigs and buds greenish or yellowish-brown, the twigs angled and buds barely aromatic; remnant leaves deltoid-shaped, and petioles flattened _____ **P. deltoides.**
    4. Twigs and buds reddish-brown, the twigs rounded and buds strongly aromatic; remnant leaves more ovate, with terete petioles _____ **P. × jackii/P. balsamifera.**

\*\* **P. alba** L. White p. Naturalized across KY and TN. Significant Threat in KY and TN. Plate 370.

**P. balsamifera** L. Balsam p. Upland woods and open sites. Few scattered records from C and E KY. FACW. Not pictured.

**P. deltoides** Bartr. ex Marshall. Eastern c. Along streams and in bottoms. Across KY and TN, but less common in far-eastern sections. FAC, FAC+. (*deltoides* = triangular). Plate 371.

**P. grandidentata** Michx. Bigtooth a. Dry to mesic woods. Chiefly E IP and E regions of KY and TN, except VR of TN. FACU–, FACU. (*grandidentata* = big toothed). Plate 372.

**P. heterophylla** L. Swamp c. Swamp forests. Chiefly CP and W IP of KY and TN. FACW+, OBL. (*heterophylla* = different-leaved). Plate 373.

\* **P. × jackii** Sarg. Balm-of-Gilead. Cultivated, persisting and spreading from root sprouts. C and E TN. (*jackii* = for John George Jack, 1861–1949, Canadian dendrologist at the Arnold Arboretum). Plate 374.

**Prunus** L. Cherry, plum. Family Rosaceae. (*Prunus* = the ancient name for a plum tree). Shrubs or trees, bark in younger stems with conspicuous horizontal banding of elongated lenticels, becoming more scaly with age; twigs slender to stout, the broken twig often with odor of bitter almond; thorny spur shoots in plums; terminal buds present in cherries and peach, but absent in plums, the buds with many imbricate scales; buds pointed to obtuse, glabrous to pubescent, sometimes clustered; leaf scars alternate, raised, rounded to elliptic, with 3 bundle scars; stipular scars minute; fruit a drupe. Remnant leaves are simple, lanceolate to ovate to orbicular, pinnately veined, and toothed, the teeth sometimes bearing a reddish gland, and the petiole often with a gland near the apex. All of these species produce edible fruits, usually maturing by late summer or early autumn. The bark is a good source of fiber and was used for a variety of medicinal purposes by Native Americans, but caution is necessary—the bark, dried leaves, twigs, and pits of fruits contain hydrogen cyanide, and are generally considered as highly poisonous.

1. Plant a small shrub, < 1 m tall at maturity _____ **P. pumila.**
1. Plants a large shrub or tree, > 1 m tall at maturity.
    2. Terminal bud present; twigs lacking thorns; trees, usually occurring singly; drupe hairy and grooved or smooth and ungrooved (peach and cherries).
        3. Buds woolly-pubescent; fruit to 8 cm thick _____ **P. persica.**

3. Buds smooth or slightly hairy; fruit to 2.5 cm thick.
    4. Twigs pubescent, at least in vicinity of buds _____ **P. mahaleb.**
    4. Twigs glabrous.
        5. Buds 2–3 mm long, blunt and clustered at twig tips and on spurs; twigs bright red; high-elevation species of TN _____ **P. pensylvanica.**
        5. Buds 4–7 mm long, sharp-pointed and usually solitary or few; twigs various; more widespread species.
            6. Spur branches usually absent; fruiting clusters, if present, elongate with > 20 fruits; native species.
                7. Mature bark nearly black, with scaly plates; buds red-brown; twigs with almond odor _____ **P. serotina.**
                7. Mature bark red-brown or darker, but with broader plates; buds light brown; twigs lacking distinctive almond odor, but may be malodorous _____**P. virginiana.**
            6. Spur branches usually present; fruiting clusters, if present, rounded with < 15 fruits; naturalized species.
                8. Tree to 20 m tall; buds glossy _____ **P. avium.**
                8. Shrub to 10 m tall; buds dull _____ **P. cerasus.**
2. Terminal bud absent; twigs often thorny; shrubs or small trees to 12 m, often occurring in colonies or thickets; drupe hairless and grooved; plums, these species indistinguishable in winter _____**Prunus americana/P. angustifolia/P. hortulana/P. mexicana/P. munsoniana/P. umbellata.**

**P. americana** Marshall. American p. Dry to mesic woods, roadsides, fencerows. Across KY and TN. Frequent. FACU–. Plate 375.

**P. angustifolia** Marshall. Chickasaw p. Dry to mesic woods, roadsides, fencerows. Across KY and TN. Frequent. (*angustifolia* = narrow-leaved). Plate 376.

\* **P. avium** (L.) L. Sweet c. Woods and stream banks. Rare in C and E KY and TN. A native of Eurasia, possibly spreading, often appearing native. (*avium* = of the birds). Not pictured.

\* **P. cerasus** L. Sour c. Disturbed areas. IP, AP of KY. Infrequent, naturalized from Europe. (*cerasus* = name for the region of Asia from which sour cherry was introduced to Rome). Plate 377.

**P. hortulana** L.H. Bailey. Hortulana p. Open woods and roadsides, uplands and lowlands. C KY, C and E TN. Infrequent. (*hortulana* = of the garden). Not pictured.

\* **P. mahaleb** L. Mahaleb c. Woods and thickets. In C KY and across TN. Infrequent, naturalized from Europe. (*mahaleb* = an Arabic vernacular name for this species). Plate 378.

**P. mexicana** S.Watson. Mexican p. Open woods and roadsides, uplands and lowlands. C and W regions of both states. Infrequent. Not pictured.

**P. munsoniana** W.Wight & Hedrick. Wild goose p. Open woods, uplands and lowlands. Chiefly C KY and C TN, few records from W and E TN. (for T.V. Munson, American botanist). Not pictured.

**P. pensylvanica** L.f. Fire c. Open woods. Mid- to upper elevations in BR of TN. FACU. Plate 379.

\* **P. persica** (L.) Batsch. Peach. Disturbed places. Across KY and TN. Infrequent, introduced from China. (*persica* = Persian). Plate 380.

! **P. pumila** L. Sand c. Fields and thickets. Known from 2 counties in S TN. Endangered in TN. (*pumila* = low, dwarf). Plate 381.

**P. serotina** Ehrh. Black c. Mesic woodlands, and also established in many disturbed areas. Across KY and TN. Frequent. FACU. (*serotina* = autumnal, of late season). Plate 382.

**P. umbellata** Elliott. Hog p. Thickets and open woodlands. C TN. Rare. (*umbellata* = umbel-like). Plate 383.

! **P. virginiana** L. Chokecherry. Woods and rocky openings. IP of KY and BR of TN. FACU. Special Concern in TN. Plate 384.

⚕/☙ **Ptelea** L. Family Rutaceae. (*Ptelea* = the ancient Greek name for elm). Shrub or small tree to 8 m tall, with thin, slightly fissured bark; twigs with wart-like lenticels, the broken twig with citrus odor; terminal bud absent; lateral buds silvery pubescent, superposed and flattened within the leaf scar; leaf scars alternate, horseshoe-shaped, surrounding the silvery buds, with 3 bundle scars; stipular scars absent; fruit a circular samara to 2.5 cm wide. The root-bark of the hop-tree was used by Native Americans, in combination with root bark from other species, for treatment of headaches (in powdered form as a snuff) and for stomach ache (in tea form). In tonic form it has been used for a wide variety of ailments, and an extract of the root-bark has been used in wound-healing. The winged fruit is bitter but not poisonous, and can be eaten. It has also been used as a substitute for hops in making beer.

**P. trifoliata** L. Common hop-tree. Dry to mesic, rocky limestone woods. IP of KY, across C and E TN. Infrequent. FAC. (*trifoliata* = having three leaflets). Plate 385.

☙/⚕/➤ **Pueraria** DC. Family Fabaceae. (for Marc Nicolas Puerari, 1765–1845, Swiss Professor of Botany at Copenhagen). Vine climbing by twining, woody mostly at stem base, often high climbing and overtopping whole sections of woodland; twigs pubescent; terminal buds absent; lateral buds in pairs at nodes, each with a large outer scale, the buds on each side of a woody peg projecting above the leaf scar; leaf scars alternate, rounded, with 3 bundle scars; stipular scars prominent, and stipules often persistent; fruit a hairy legume, but rarely fruiting. Kudzu has multiple uses (known as one of the 50 fundamental herbs in China). The roots (high in starch) and the young shoots and leaves (very nutritious) can be cooked and eaten. The pith of the stems can be removed and repeatedly washed and dried and made into a flour. Extracts of the roots and flowers suppress the desire for alcohol and have been used for treating alcohol abuse. It is used for treating migraine headaches, for increasing blood flow through the arteries, and for many other ailments. The stems are also a good source of fiber for cordage

** **P. montana** (Lour.) Merr. var. **lobata** (Willd.) Maesen & S.M.Almeida. Kudzu. Thickets and woodlands. Infrequent, but often locally abundant, across KY and TN, progressively less frequent northward. Naturalized from Japan, cultivated for forage and erosion-control, but now widely established and a troublesome weed across the SE U.S. Severe Threat in KY and TN. (*montana* = of mountains). Plates 386 and 387.

☠ **Pyrularia** Michx. Family Santalaceae. (*Pyrularia* = little pear). Shrub to 3 m tall, parasitic on deciduous trees, with smooth gray bark; twigs with spongy-porous pith; terminal buds usually present, imbricate-scaly, ovoid, to 12 mm long, the scales green with brown or purple margins; leaf scars alternate, half-round, with 3 bundle scars; the branch scars rounded with many bundle scars in an ellipse; stipular scars absent; fruit a green, oily drupe to 2.5 cm long, poisonous. Remnant leaves simple, lanceolate, pinnately veined, and entire.

**P. pubera** Michx. Buffalo-nut. Mixed mesophytic forests. E KY and E TN. Infrequent. UPL, FACU+. (*pubis* = hairy). Plate 388.

☙ **Pyrus** L. Pear. Family Rosaceae. (*Pyrus* = the ancient name for pear tree). Trees to 15 m tall, with gray-scaly bark; twigs sometimes thorn-tipped; terminal buds present, scaly-imbricate; leaf scars alternate, crescent-shaped, with 3 bundle scars; stipular scars absent; fruit a pome, with gritty flesh. Remnant leaves are simple, ovate to rounded, pinnately veined, and toothed. It is the common pear that produces the edible fruit; the fruits of the Bradford (or Callery) pear are usually much smaller and bitter.

1. Buds and twigs glabrous; buds to 8 mm long _____**P. communis.**
1. Buds and twigs pubescent; buds > 8 mm long _____**P. calleryana.**

\*\* **P. calleryana** Decne. Bradford p. Widely planted, native of China, and increasingly becoming naturalized and a serious invasive species in KY and TN. (*calleryana* = for Joseph Callery, 1810–1862, missionary and botanist in Korea and China). Severe Threat in KY and an Alert Species in TN. Plates 389, 390, and 391.

\* **P. communis** L. Common p. Disturbed places. Across KY and TN. Infrequent, introduced from Eurasia. (*communis* = growing in clumps, gregarious, common). Plate 392.

**Quercus** L. Oak. Family Fagaceae. (*Quercus* = the old Latin name for oak). Mostly trees, the leaves often tardily deciduous and persistent in winter; twigs with pith 5-angled in cross-section, with clustered end buds; terminal buds present, imbricate-scaly; leaf scars alternate, half-round, with many bundle scars; stipular scars small, the stipules sometimes persistent; fruit a nut with a cup-like involucre. Remnant leaves are often persistent, as described in the key to species. Acorns of the white oaks and chestnuts oaks (see below) are relatively sweet, have low tannin amounts, and require no leaching; they can be collected as they turn brown in the fall, stored in cool dry place, and used for grits (dried and coarsely ground) or ground finely and redried into flour. The acorns of the red oaks have high tannins and need to be chopped or ground and leached repeatedly (suspended with a cloth in boiling water) until water remains clear, and then used as described above for white oak acorns. Extracts from the bark were used by Native Americans for a wide variety of ailments, but primarily for healing wounds, and there are ongoing investigations into possible anti-cancer effects. Tannins can be toxic at high concentrations, so caution should be used; long-term ingestion of leaves, twigs, or high-tannin acorns can lead to a variety of metabolic problems. The sturdy young branches are among the best for constructing arrow shafts, and the inner bark is a good source of fiber.

1. Immature acorns absent from twigs in winter (acorns mature in one year); inner surface of mature acorns smooth; cup scales thickened or knobby; leaves lacking bristle tips; bark various, but not black and blocky or tightly ridged and furrowed (in most species either gray or flaky or both, or with deep furrows in *Q. montana*); white oaks and chestnut oaks.
    2. Bark dark and deeply furrowed with wide ridges; buds and twigs orange-brown, the buds slender and acute, 6–10 mm long; acorn cup sharply and thinly edged (not frilly) around the rim, covering about ⅓ of nut; nut elliptic, 2.5–3.5 cm long; leaves with regular rounded teeth; upland species of C and E KY and TN _____**Q. montana.**
    2. Bark gray or flaky or both; buds and acorns and habitat various, but not in above combination; leaves various, but if regularly toothed, then mucronate tips present on teeth.
        3. Twigs with corky growth, often winged; bark gray-flaky; bristle-like stipules often present around terminal buds, these to 6 mm long; acorns large, often > 4 cm long and wide, the cup covering about ½ of the nut, with a distinctive frilly edge; leaves shallowly lobed toward apex, more deeply lobed toward base; habitat various, often in uplands _____**Q. macrocarpa.**
        3. Twigs lacking corky growth; buds usually < 5 mm long (except in *Q. michauxii*); acorns and leaves otherwise; habitat various.
            4. Buds globose to obtuse; leaves various, but margins typically shallowly to deeply lobed.
                5. Bark ridged and scaly; twigs yellowish and pubescent; buds reddish-brown, often pubescent; stipules sometimes persistent; leaves symmetrically lobed, usually with 2 large lobes at middle; nut to 2 cm long, with cup covering about ⅓ of nut; habitat various, often in uplands or sandy site _____**Q. stellata.**

5. Bark gray and flaky; twigs, bud, and pubescent pattern otherwise, the twigs usually glabrous or nearly so; stipules various; leaves lacking large middle lobes (except sometimes in *Q. margaretta*); acorns various; habitat various.
   6. Cup < 1 cm deep, covering 1/3 of nut, stipules absent; leaves regularly and often deeply lobed; plants of widespread distribution, mostly of uplands_____ **Q. alba.**
   6. Cup > 1 cm deep, covering > 1/3 of nut; stipules often persistent; leaves irregularly lobed to shallowly toothed; plants of wetlands.
      7. Shrub or small tree to 12 m tall; cup scales flat; lower leaf surfaces pubescent with short erect hairs, lacking stellate hairs; restricted to southwestern corner of TN _____**Q. margaretta.**
      7. Tall trees to 30 m tall; cup scales awned or keeled; lower leaf surfaces pubescent with stellate hairs or with both stellate and erect hairs; widespread species.
         8. Bud scales glabrous; cup < 15 mm deep, covering < 3/4 of nut, separating from the nut at maturity; acorn stalk 2–6 cm long_____ **Q. bicolor.**
         8. Buds scales pubescent or ciliate; cup > 15 mm deep, covering > 3/4 of nut, remaining with the nut at maturity; acorn sessile or nearly so _____ **Q. lyrata.**
  4. Buds acute, conical; leaves shallowly toothed.
   9. Buds 5–7 mm long, reddish; acorns 2.5–3.5 cm long, the cup > 2 cm wide; plants of lowlands of western portions of KY and TN_____ **Q. michauxii.**
   9. Buds to 4 mm long, brownish; acorns 1–2 cm long, the cup < 2 cm wide; plants of upland regions of C and E KY and TN.
      10. Shrubs, mostly < 5 m tall, multistemmed and forming colonies; secondary leaf veins < 10 _____**Q. prinoides.**
      10. Trees to 30 m tall, the stems usually solitary; secondary leaf veins > 10 _____**Q. muhlenbergii.**
1. Immature acorns present on twigs (acorns mature in 2 years); inner surface of mature acorns pubescent; cup scales flattened on the back; leaves with bristle tips; bark various, but not gray and flaky; red oaks, black oaks, and willow oaks.
  11. Bark blackish and rough-blocky (platy in *Q. pagoda*), lacking shiny strips on upper branches; buds and twigs pubescent (glabrate in *Q. velutina*); leaves shallowly to deeply lobed, or 3-lobed at apex; cup scales loosely covering the cup, usually thick and blunt-tipped.
     12. Buds to 12 mm long, distinctly angled in cross-section; nuts to 2 cm long.
        13. Twigs glabrate; inner bark yellow; leaves 5–9 lobed, on petioles > 2 cm long ___**Q. velutina.**
        13. Twigs pubescent; inner bark orange; leaves 3-lobed at apex, on petioles to 2 cm long _____ **Q. marilandica.**
     12. Buds to 6 mm long, rounded or slightly angled; nuts to 1.5 cm long.
        14. Bark platy, similar to that of black cherry; buds angled; leaves regularly 5–11 lobed, tapering to petiole; plants restricted to lowlands of W KY and TN _____ **Q. pagoda.**
        14. Bark blocky; buds rounded; leaves variable, up to 7 lobed, often with terminal lobe elongate or with only 3 short apical lobes; plants widespread across the region, often in uplands _____ **Q. falcata.**
  11. Bark grayish or brownish, usually thinly ridged and furrowed on trunk, and frequently with shiny strips above on larger branches; twigs glabrous, buds various; leaves various; cup scales tightly covering the cup, the scales often acute or acuminate-tipped.
    15. Leaves all entire and unlobed.
      16. Twigs very slender, < 2 mm thick, bud rounded; leaves < 2 cm wide, glabrous below, and petiole < 5 mm long; nuts < 1.5 cm long _____**Q. phellos.**
      16. Twigs thicker, to 2.5 mm thick, buds often angled; leaves > 2 cm wide, pubescent below, and petioles > 5 mm long; nuts to 2 cm long_____ **Q. imbricaria.**

15. Leaves lobed (some leaves may be unlobed in *Q. nigra*, but others are lobed apically).
    17. Buds white-tipped with pale pubescence; cup top- or bowl-shaped, appearing shiny with varnished appearance, covering up to ½ of nut, the nut tip with concentric rings\_\_ **Q. coccinea.**
    17. Buds various, but of uniform coloration; cup more saucer- or bowl-shaped, dull colored, covering about ⅓ of the nut.
        18. Buds both angled and pubescent; cup covering ¼ of nut, the nut < 1.5 cm long; leaves with 3 or fewer terminal lobes _____ **Q. nigra.**
        18. Buds otherwise; cup various, the nut larger except sometimes in *Q. palustris*; leaves with 5 or more lateral lobes.
            19. Leaves with 5–7 lobes; nut to 1.5 cm long and wide; twigs and buds smooth; lower branches of tree typically drooping_____ **Q. palustris.**
            19. Leaves with 5–11 lobes; nuts to 3 cm long and wide; twigs and buds various, the latter often angled or with some pubescence; branches various.
                20. Inner surface of acorn cup uniformly pubescent; buds gray-brown, unangled, slightly pubescent or ciliate; distributed only in western portions of KY and TN _____ **Q. texana.**
                20. Inner surface of acorn cup glabrous or with ring of hairs around scar; buds otherwise; species distributed across most of the region.
                    21. Buds gray and angled, glabrous; twigs greenish or grayish; cup more bowl-shaped, the scales with acuminate tips; leaf sinus narrow, tending to close toward tips _____ **Q. shumardii.**
                    21. Buds and twigs red-brown, the buds sometimes slightly red-ciliate; cup more saucer-shaped, the scales with acute tips; leaf sinuses wide and spreading _____ **Q. rubra.**

**Q. alba** L. White o. Mesic slopes to dry uplands. Across KY and TN. Frequent. FACU–, FACU. Plates 393 and 394.

**Q. bicolor** Willd. Swamp white o. Swamp forests and wet woods. Chiefly C KY and TN, rare in E and W portions of both states. Infrequent. FACW+. (*bicolor* = of two colors). Plates 395 and 396.

**Q. coccinea** Münchh. Scarlet o. Acidic upland woods. Across KY and TN, but less common in central and far-western portions of the region. Frequent. (*coccinea* = crimson, scarlet). Plates 397 and 398.

**Q. falcata** Michx. Southern Red o., Spanish o. Acidic upland to lowland woods. Across KY and TN, in all regions except N KY. Frequent. FACU–. (*falcate* = sickle-shaped). There is a historical record of *Quercus ilicifolia* Wang., the bear oak, for KY—it is similar to *Q. falcata* in small buds and hairy twigs, but is more shrublike, with leaves < 10 cm long. Plates 399 and 400.

**Q. imbricaria** Michx. Shingle o. Dry to mesic woods, chiefly calcareous sites. Across KY and TN, but much less frequent in E and W portions of both states. Frequent. FAC, FAC–. (*imbricaria* = overlapping). Plates 401 and 402.

**Q. lyrata** Walter. Overcup o. Swamp forests. Chiefly CP and IP. Infrequent. OBL. (*lyrata* = shaped like a lyre). Plates 403 and 404.

**Q. macrocarpa** Michx. Bur o. Mesic to dry woodlands, usually open and savanna-like. Chiefly C KY and TN, few records westward. Infrequent. FAC–, FAC. (*macrocarpa* = large fruited). Plates 405 and 406.

! **Q. margaretta** (Ashe) Ashe. Dwarf post o. Sandy soils, open woodlands. Known only from Fayette County, TN. Rare. Special Concern in TN. (*margaretta* = pearly). Plate 407.

**Q. marilandica** Münchh. Blackjack o. Rocky upland woods. Across both states, but rare to absent in W KY, W TN, N KY, and C TN. Infrequent. (*marilandica* = from the Maryland region, U.S.). Plates 408 and 409.

**Q. michauxii** Nutt. Swamp chestnut o., basket o., cow o. Swamp forests. Across TN, but chiefly CP and S IP in KY. Infrequent. FACW, FACW–. (*michauxii* = for Andre Michaux, 1746–1803, French botanist who explored eastern North America for 11 years, and wrote the first guide to the North American flora based entirely on the author's own botanical studies). Plates 410 and 411.

**Q. montana** Willd. Chestnut o. Dry upland woods, often dominant on ridgetops. Chiefly in E KY, absent W of Tennessee River; across C and E TN. Frequent. UPL. Formerly known as *Q. prinus* L. (*montana* = of mountains). Plates 412 and 413.

**Q. muhlenbergii** Engelm. Chinkapin o. Dry calcareous uplands woods. Across KY and TN. Frequent. (*muhlenbergii* = for Henri Ludwig Mühlenberg, 1756–1817, of Pennsylvania). Plates 414 and 415.

! **Q. nigra** L. Water o. Swamp forests. In CP and S IP of TN, much rarer in KY, known only from few counties in CP and IP. FAC. Threatened in KY. Plates 416 and 417.

**Q. pagoda** Raf. Cherrybark o. Bottomland hardwoods. Chiefly CP of both states, less common in IP, and scattered E in TN. Infrequent. FACW, FAC+. (*pagoda* = with the habit of a pagoda, pyramidal). Plates 418 and 419.

**Q. palustris** Münchh. Pin o. Swamp forests and wet woods. Chiefly W and C KY, and across TN except far E TN. Widely planted. FACW. (*palustris* = of swampy ground). Plates 420 and 421.

**Q. phellos** L. Willow o. Swamp forests and wet woods. Chiefly CP and S IP of KY; across TN except the far-eastern portions. Infrequent. FAC+, FACW–. (*phellos* = cork). Plates 422 and 423.

**Q. prinoides** Willd. Dwarf chinkapin o. Upland woods and bluffs, especially sandy sites. Infrequent. C and E TN. (*prinoides* = similar to chestnut oak). Plate 424.

**Q. rubra** L. Northern red o. Mesic woodlands. Across KY and TN. Frequent. FACU–, FACU. Plates 425 and 426.

**Q. shumardii** Buckley. Shumard o. Mesic limestone woods. Chiefly IP, especially limestone regions, much less frequent to E and W in both states. Frequent. FAC+, FACW–. (*shumardii* = for B.F. Shumard, 1820–1869, Texas geologist). Plates 427 and 428.

**Q. stellata** Wangenh. Post o. Dry to mesic, often rocky woods. Across KY and TN. Frequent. UPL, FACU. (*stellata* = with spreading rays, stellate, star like). Plates 429 and 430.

! **Q. texana** Buckley. Nuttall's o., Texas red o. Bottomland hardwoods. CP, less common in KY. Threatened in KY. OBL (in TN). Formerly known as *Q. nuttallii* E.J.Palmer. Plates 431 and 432.

**Q. velutina** Lam. Black o. Dry to mesic, primarily upland woods. Across KY and TN. Frequent. (*velutina* = finely velvety). Plates 433 and 434.

**Rhamnus** L. Buckthorn. Family Rhamnaceae. (*rhamnus* = an ancient name for various prickly shrubs). Shrubs with smooth gray outer bark and yellow inner bark; terminal bud lacking; buds scaly, to 5 mm long, the scales dark-edged with white-ciliate margins; leaf scars opposite or alternate, crescent-shaped with 3 bundle scars, or sometimes these fused into a line; stipular scars tiny; fruit berrylike, red to black. Remnant leaves simple, lanceolate to elliptic, pinnately veined, and toothed. Ingestion of the bitter fruit, bark, or leaves can stimulate a strong laxative effect.

1. Twigs thorny; leaf scars opposite or subopposite.
    2. Twigs glabrous; remnant leaves shiny above _____ R. davurica.
    2. Twigs hairy; remnant leaves dull above _____ R. cathartica.
1. Twigs unarmed; leaf scars alternate.
    3. Erect shrub to 2 m tall; twigs pubescent; remnant leaves sharply toothed _____ R. lanceolata.
    3. Sprawling shrub < 1 m tall; twigs glabrous; remnant leaves bluntly toothed _____ R. alnifolia.

! **R. alnifolia** L'Hér. Alderleaf b. Moist limestone woods and seeps. VR of TN. OBL. Endangered in TN. (*alnifolia* = *Alnus*-leaved). Plate 435.

* **R. cathartica** L. European b. Thickets and woods. Infrequent to rare escape in KY and TN. Naturalized from Europe. UPL (in KY). (*cathartica* = cleansing, purging). Not pictured.

* **R. davurica** Pall. Dahurian b. Thickets and woods. Infrequent to rare escape in KY and TN. Naturalized from China. [*R. citrifolia* (Weston) W.J.Hess & Stearn.] (*davurica* = from Dauria, NE Asia). Plate 436.

**R. lanceolata** Pursh. Lanceleaf b. Dry to mesic woods. IP, chiefly BG of KY and CB of TN. Infrequent. (*lanceolatus* = narrowed and tapered at both ends). Plate 437.

☙ **Rhododendron** L. Azalea, laurel, rhododendron. Family Ericaceae. (*Rhododendron* = rose tree). Shrubs, some species with evergreen leaves; twigs often with whorls of branches near tips; terminal buds present, scaly-imbricate, with upper lateral buds clustered near tips, the flower buds larger; buds ovoid, the scale tips often rounded-mucronate or abruptly-acuminate; leaf scars alternate, shield-shaped to linear, with single bundle scar; stipular scars absent; fruit a cylindric 5-chambered capsule, often present in winter. In both evergreen and deciduous species the leaves are simple, oblanceolate to elliptic, pinnately veined, and entire. All parts are highly toxic (see note under *Kalmia*).

1. Leaves evergreen.
    2. Twigs and lower leaf surfaces dotted with brown scales_____ R. minus.
    2. Twigs and lower leaf surfaces lacking brown scales.
        3. Leaves rounded at the apex; calyx lobes < 2 mm long; capsules eglandular _____ R. catawbiense.
        3. Leaves pointed at the apex; calyx lobes 2–6 mm long; capsules glandular _____ R. maximum.
1. Leaves deciduous.
    4. Twigs and buds glabrous (bud scales may be ciliate) _____ R. arborescens.
    4. Twigs pubescent, buds glabrous or pubescent.
        5. Capsules with glandular trichomes.
            6. Twigs downy-pubescent _____ R. prinophyllum.
            6. Twigs bristly-pubescent _____ R. viscosum.
        5. Capsules lacking glandular trichomes.
            7. Bud scales surfaces densely pubescent; capsules densely pubescent _____ R. canescens.
            7. Bud scale surfaces glabrous or slightly pubescent; capsules various.
                8. Buds scale margins glandular_____ R. calendulaceum/R. cumberlandense.
                8. Bud scale margins ciliate with eglandular trichomes.
                    9. Capsules widest at base _____ R. periclymenoides.
                    9. Capsules widest at middle_____ R. alabamense.

**R. alabamense** Rehder. Alabama a. Dry open woods. C TN (CB and AP). Plates 438 and 439.

**R. arborescens** (Pursh) Torr. Smooth a., Sweet a. Swamp forests and stream banks. S AP of KY and from E IP to BR in TN. Infrequent. FAC, FACW–. (*arborescens* = becoming or tending to be of tree-like dimensions). Plates 440 and 441.

**R. calendulaceum** (Michx.) Torr. Flame a. Dry to mesic forests. AP of KY and VR and BR of TN. Infrequent. (*calendulaceum* = with golden flower heads). Plates 442 and 443.

! **R. canescens** (Michx.) Sweet. Southern pinxter a. Swamp forest. Rare in CP of KY, more common across TN, especially in IP; Endangered in KY. FACW, FACW–. (*canescens* = turning hoary-white, with off-white pubescence). Plates 444 and 445.

**R. catawbiense** Michx. Rosebay l. Rocky open woods and meadows. Rare in far E KY, and infrequent from AP to BR in TN. (*catawbiense* = from the area of the Northern American Indian Catawba tribe, from the Catawba River area, North Carolina). Plates 446 and 447.

**R. cumberlandense** E.L.Braun. Cumberland a. Dry to mesic oak and oak-pine forest. AP of KY and AP to BR in TN. Infrequent. [*R. bakeri* (Lemon & McKay) Hume]. (*cumberlandense* = from the Cumberlands). Plates 448 and 449.

**R. maximum** L. Great l. Mixed mesophytic forest. AP of KY; AP to BR in TN, with few outlying populations in IP of both states. Frequent. FAC, FAC–. (*maximum* = largest, greatest). Plates 450 and 451.

**R. minus** Michx. Piedmont r. Steep wooded or open cliffs and stream banks. VR and BR of TN. Infrequent. (*minus* = small, less). Plate 452.

**R. periclymenoides** (Michx.) Shinners. Pink a., pinxterbloom. Swamp forests and along streams. AP of KY and in IP (except CB), AP, VR, and BR in TN. Frequent. FAC. (*periclymenoides*, a twining plant, from the writings of Dioscorides). Plates 453 and 454.

**R. prinophyllum** (Small) Millais. Rosebud a. Dry to mesic woodlands and along streams. AP of KY. Rare. FAC. (*prinophyllum* = having oak-like leaves). Not pictured.

**R. viscosum** (L.) Torr. Swamp a. Shrub balds. BR of TN. FACW+. (*viscosum* = sticky, viscid). Plate 455.

**Rhodotypos** Siebold & Zucch. Family Rosaceae. (*Rhodotypos* = rose type). Shrub to 2 m tall; twig tips dying in winter; terminal buds absent; lateral buds imbricate-scaly, the scales dark-tipped, and collateral buds often present; leaf scars opposite, connected by a line, crescent- to triangular-shaped, with 3 bundle scars; stipular scars or dried stipules present; fruit a black drupe, to 8 mm long, subtended by persistent sepals. Remnant leaves simple, ovate and long pointed, pinnately veined, and coarsely toothed. The fruit is considered highly toxic.

* **R. scandens** (Thunb.) Makino. Jetbead. Disturbed places. Scattered records of naturalized plants across C and E KY and TN. Rare, cultivated, a native of E Asia. (*scandens* = climbing). Plates 456 and 457.

**Rhus** L. Sumac. Family Anacardiaceae. (*Rhus* = from an ancient Greek name for a sumac). Shrubs or small trees (to 10 m tall) with resinous or milky sap; twigs brittle, with large pith; terminal buds absent; lateral buds small, hairy, and naked; leaf scars alternate, half-round to horseshoe-shaped, bundle scars > 3; stipular scars absent; fruit a red drupe, in panicles, often persistent into winter. All of these species produce edible fruits—the drupes are lemony tasting, and typically used to prepare a beverage by steeping them in hot water, straining out the hairs, and sweetening (some people may have an allergic reaction). This brew may also be helpful for alleviating a sore throat.

1. Plant a low shrub, to 2 m tall; twigs with pungent aroma when broken; catkin-like flower spikes present; leaf scars rounded_____ **R. aromatica.**
1. Plant a taller shrub or small tree; twigs lacking pungent aroma; catkin-like structures absent; leaf scars otherwise.
   2. Leaf scars half-round, not encircling the bud; twigs < 1 cm thick; remnant panicles drooping_____
   _____ **R. copallinum.**
   2. Leaf scars nearly encircling the buds; twigs > 1 cm thick; remnant panicles erect.
      3. Twigs glabrous _____**R. glabra.**
      3. Twigs velvety-pubescent_____ **R. typhina.**

**R. aromatica** Aiton. Fragrant s. Rocky limestone woodlands. Chiefly limestone regions of IP of KY and TN, rare to absent in W and E sections. Frequent. (*aromaticus* = fragrant). Plate 458.

**R. copallinum** L. Winged s. Open woodlands, roadsides and fields. Across KY and TN. Frequent. (*copallina* = from a Mexican name, *kopalli*, yielding copal-gum). Plate 459.

**R. glabra** L. Smooth s. Open woodlands, fields and roadsides. Across KY and TN. Frequent. (*glabra* = smooth, bald). Plates 460 and 461.

**R. typhina** L. Staghorn s. Open woodlands, fields, and roadsides. IP and AP of KY, IP to BR in TN. Infrequent. (*typhina* = bulrush-like). Plates 462 and 463.

**Ribes** L. Currant, gooseberry. Family Grossulariaceae. (*Ribes* = from *ribes*, the Danish word for currants). Shrubs, often with shreddy bark; twigs with spines attached at base of leaf scars or spines absent, ridged from the nodes; pith continuous to spongy or porous in older twigs; terminal buds present, few-scaled, often lanceolate; buds imbricate-scaly; leaf scars alternate, very narrow, with 3 bundle scars; stipular scars absent; fruit a berry, but rarely present in winter. Remnant leaves simple, palmately lobed and veined, and toothed. All species have edible berries, but they vary in sweetness and palatability. They can be eaten fresh, used for pies and jellies, or dried for later use. The species most likely to be encountered in the region are *R. cynosbati* and *R. missouriense* in the IP, and *R. rotundifolium* and *R. glandulosum* in the higher elevations of E TN.

1. Nodal spines absent; berries black (currants).
   2. Shrubs to 1 m tall; twigs and buds glabrous; inner bark with skunk-like odor; fruit bristly _____
   _____ **R. glandulosum.**
   2. Shrubs > 1 m tall; twigs and buds pubescent; inner bark lacking skunk-like odor; fruit smooth.
      3. Bark not shreddy, the inner bark lacking odor_____ **R. aureum.**
      3. Bark shreddy, the inner bark aromatic _____ **R. americanum.**
1. Nodal spines present; berries green, red, or purple (gooseberries).
   4. Berries with persistent prickles _____**R. cynosbati.**
   4. Berries glabrous or pubescent.
      5. Nodal spines 1–2 cm long _____ **R. missouriense.**
      5. Nodal spines < 1 cm long.
         6. Twigs and nodal spines reddish _____**R. curvatum.**
         6. Twigs and nodal spines grayish_____**R. rotundifolium.**

! **R. americanum** Mill. Wild Black c. Hills and slopes, mesic forest. IP and AP of KY, and VR of TN. Rare. Threatened in KY. FACW, FAC. Not pictured.

! **R. aureum** Pursh var. **villosum** DC. Buffalo c. River bluffs of W IP of TN. Threatened in TN. (*R. odoratum* H.L. Wendl.). (*aureum* = golden-yellow). Plate 464.

! **R. curvatum** Small. Granite g. Rocky woodlands. AP of TN. Rare. Threatened in TN. (*curvatum* = curved). Plate 465 and 466.

**R. cynosbati** L. Eastern prickly g. Rocky slopes; mesic to dry forest. IP and AP of KY, E TN. Infrequent. (*cynosbati* = dog thornbush). Plate 467.

**R. glandulosum** Grauer. Skunk c. High-elevation woods and openings. BR of TN. Infrequent. FACW. (*glandulosum* = glandular). Plate 468.

! **R. missouriense** Nutt. ex Torr. & A.Gray. Missouri g. Slopes and ledges; dry to mesic woods. IP of KY and TN. Infrequent. Special Concern in TN. Plate 469.

**R. rotundifolium** Michx. Appalachian g. High-elevation woods. BR of TN. Infrequent. (*rotundifolium* = round-leaved). Plate 470.

**Robinia** L. Locust. Family Fabaceae. (for Jean Robin, 1550–1629, and Vesparian Robin, 1579–1600, herbalists and gardeners for Henry VI of France). Shrubs or trees, often rapidly spreading by root sprouts, the bark in older trees thickly ridged and furrowed; twigs zigzag, slightly angled; terminal buds absent; lateral buds sunken, superposed, and downy-hairy, hidden by a membrane over the leaf scar (which splits as buds enlarge); leaf scars alternate, triangular, with 3 bundle scars, but these obscured; stipules modified into paired nodal spines (especially in juvenile specimens of *R. pseudoacacia*), but twigs often spineless in mature individuals; fruit a legume to 10 cm long, lacking pulp, and splitting at maturity, revealing the black seeds. All parts, but especially the inner bark and seeds of these species, contain toxic compounds and should not be ingested. The sturdy young branches are among the best for constructing bows and arrows.

1. Twigs glabrous; trees _____ **R. pseudoacacia.**
1. Twigs bristly or with glandular hairs; shrubs or small trees _____ **R. hispida.**

**R. hispida** L. Bristly l. Dry to mesic forest; disturbed places. Across KY and TN. Infrequent. (*hispida* = bristly hairy). Wofford and Chester (2002) noted, with reference to *R. hispida*, that "…Numerous named and perhaps some unnamed, mostly sterile and clonal hybrids and introgressants occur in TN, …They appear after disturbance and are often short-lived; to attempt to put them into a key format would be futile." Plate 471.

**R. pseudoacacia** L. Black l. Disturbed sites, roadsides, mesic woods. Across KY and TN. Abundant. FACU–, UPL. (*pseudoacacia* = false *Acacia*). Plates 472, 473, and 474.

**Rosa** L. Rose. Family Rosaceae. (*Rosa* = the Latin name for various roses). Shrubs, the branches climbing or scrambling, often with flat-based prickles, these scattered or paired; terminal buds present, imbricate-scaly, with 3 or 4 visible scales; leaves evergreen in some species, and often persistent, pinnately compound with 3 to 9 leaflets; leaf scars alternate, very narrow, and about half-encircling the twig, with 3 bundle scars; stipular scars absent; fruit of achenes enclosed in a red or orange fleshy hypanthium (a hip). The hypanthium is edible, and very high in vitamin C. Preparation requires removing the dried sepals from the fruit top and the achenes from inside the hip; the fleshy red cuplike hypanthium can be eaten raw, or dried and ground and used as a flavoring or tea. The stems can be utilized as arrow shafts.

1. Plants climbing or scrambling; styles united into a column protruding from hypanthium; hips lacking stalked glands (except in *R. setigera*).
    2. Leaves evergreen, with 5–9 glossy leaflets; stipules jagged dentate _____ **R. wichuraiana.**
    2. Leaves deciduous, or if present, not glossy; stipules otherwise.
        3. Stipules entire or nearly so; pedicels and hips stipitate-glandular; remnant leaves with leaflets 3 or 5___
        _____ **R. setigera.**
        3. Stipules pectinate; pedicels and hips usually glabrous; remnant leaves with leaflets 7 or 9_____
        _____ **R. multiflora.**

1. Plants with stems erect and shrubby; styles distinct, shortly exserted or enclosed by hypanthium; hips pubescent with stalked glands.
    4. Hypanthium about 1 mm wide, the styles shortly exserted; sepals dissimilar, the outer ones pinnatifid; remnant leaves red-dotted below _____ **R. eglanteria.**
    4. Hypanthium opening 2–4 mm wide, plugged by the styles; sepals all similar; remnant leaves lacking red dots.
        5. Nodal prickles recurved, flattened just above the broadened base.
            6. Leaf teeth finely serrate, with 9–11 teeth/cm; lowland species _____**R. palustris.**
            6. Leaf teeth coarsely toothed, with 5–7 teeth/cm; chiefly upland species _____ **R. virginiana.**
        5. Nodal prickles straight, rounded just above slender base (internodal prickles sometimes broad-based); leaves coarsely toothed with 5–7 teeth/cm; upland species _____**R. carolina**

**R. carolina** L. Pasture r. Open woodlands, pastures, thickets. Across KY and TN. Frequent. UPL, FACU. Plates 475 and 476.

\* **R. eglanteria** L. Eglantine r. Disturbed places. Rare escape across both states. Introduced from Europe. (*eglanteria* = French name for rose-hips). (*R. rubiginosa* L.). Not pictured.

\*\* **R. multiflora** Thunb. ex Murray. Multiflora r. Disturbed places, fields and woodlands. Across KY and TN. Frequent, naturalized from Asia, originally introduced in the state for wildlife food and habitat. Significant Threat in KY and TN. FACU. (*multiflora* = many flowered). Plates 477 and 478.

**R. palustris** Marshall. Swamp r. Swamps and marshes. Across KY and TN. Infrequent. OBL. (*palustris* = of swampy ground). Plates 479 and 480.

**R. setigera** Michx. Prairie r. Dry to mesic open woods, thickets, fencerows, roadsides. Across KY and TN. Frequent. FACU. (*setigera* = bearing bristles). Plates 481 and 482.

! **R. virginiana** Mill. Virginia r. Open limestone woods, thickets, and roadsides. Rare across KY and TN. FAC. Special Concern in TN. Plates 483 and 484.

\* **R. wichuraiana** Crepin. Memorial r. Occasionally escaping in KY and TN, native of Asia. (*wichuraiana* = for Max Ernst Wichura, 1817–1866, German botanist). Not pictured.

**Rubus** L. Blackberry, dewberry, raspberry. Family Rosaceae. (*Rubus* = the ancient Latin name for brambles). Shrubs, usually prickly or bristly or both, with stems arching, trailing, scrambling, or erect; terminal buds absent; the lateral buds ovoid to lanceolate, often superposed, imbricate-scaly; leaves alternate, the petiole incompletely deciduous, the basal portion remaining as a "petiolar stump," and the leaf scars and bundle scars hidden under petiole base (persistent leaves with 3 or 5 leaflets); stipules present, often persistent on petiolar stalk; fruit an aggregate of drupelets, absent in winter. These fleshy and often delicious fruits ripen in midsummer, and are a primary source of food for many kinds of wildlife, including songbirds, gamebirds, and mammals.

1. Stems trailing, the erect branches < 20 cm tall (dewberries).
    2. Stems prickly, not hispid _____ **R. flagellaris.**
    2. Stems prickly and bristly-hispid.
        3. Bristles gland-tipped_____**R. trivialis.**
        3. Bristles pointed, lacking glands_____**R. hispidus.**
1. Stems arching to scrambling or shrubby, the branches > 20 cm long.
    4. Bark shreddy and twigs glandular hairy but lacking prickles; receptacle flat; leaves simple _____ **R. odoratus.**
    4. Bark and twigs and prickles not in above combination (unarmed only in the smooth-barked *R. canadensis*); receptable conical; leaves compound.

5. Stems either white-glaucous OR strongly bristly; fleshy fruit separating from receptacle (raspberries); leaflets white below.
   6. Stems whitish-glaucous, lacking bristles or dense pubescence _____ **R. occidentalis.**
   6. Stems green, densely bristly or pubescent.
      7. Stems with stiff glandular bristles; calyx gray-tomentose _____ **R. idaeus.**
      7. Stems densely shaggy with purplish glandular trichomes; calyx shaggy with stipitate trichomes _____ **R. phoenicolasius.**
5. Stems not as above; fleshy fruit remaining attached to receptacle. (blackberries); leaflets green or white below.
   8. Fruiting branches with few, slender prickles; leaflets green below; stems erect or arching upward; fruiting clusters racemiform or fruits solitary (natives).
      9. Pedicels with glandular trichomes _____ **R. allegheniensis.**
      9. Pedicels lacking glands.
         10. Stems unarmed, or with a few scattered prickles _____ **R. canadensis.**
         10. Stems conspicuously armed with stout prickles.
            11. Stems usually angled; stem prickles straight; terminal leaflet, if present, cuneate at base _____ **R. argutus.**
            11. Stems more rounded; stem prickles hooked; terminal leaflet rounded or cordate at base _____ **R. pensilvanicus.**
   8. Fruiting branches with many stout prickles; leaflets white below; flowering clusters cymose-paniculate; stems scrambling horizontally (exotics) _____ **R. bifrons/R. longii.**

**R. allegheniensis** Porter. Allegheny b. Thickets, open woodland and disturbed habitats. Chiefly IP and AP of KY and TN, also in BR of TN. Frequent. FACU–, UPL. (*allegheniensis* = from the Alleghany mountains). Plate 485.

**R. argutus** Link. Southern b. Mesic to wet woods, thickets, and various disturbed habitats. Across KY and TN. Frequent. FACU, FACU+. (*argutus* = sharply toothed or notched). (*R. betulifolius* Small). Plates 486 and 487.

\* **R. bifrons** Vest ex Tratt. Himalayan b. Disturbed places. Across TN but only CP of KY. Rare, escaped from cultivation, introduced from Europe. (*bifrons* = having a double garland of leaves, two-boughed). Plates 488 and 489.

! **R. canadensis** L. Smooth b. High-elevation forests and thickets. BR of TN and in SE AP of KY. Endangered in KY. Plate 490.

**R. flagellaris** Willd. Northern d. Dry to mesic open woods, various disturbed habitats. Across KY and TN. Infrequent. UPL. [*R. baileyanus* Britton, *R. deamii* L.H.Bailey, *R. enslenii* Tratt., *R. depavitus* L.H.Bailey, *R. fecundus* L.H.Bailey, *R. felix* L.H.Bailey, *R. indianensis* L.H.Bailey, *R. invisus* (L.H.Bailey) Britton, *R. kentuckiensis* L.H.Bailey, *R. leviculus* L.H.Bailey, *R. meracus* L.H.Bailey, *R. roribaccus* (L.H.Bailey) Britton, *R. whartoniae* L.H.Bailey]. (*flagellaris* = with long thin shoots, whip-like). Plate 491.

**R. hispidus** L. Swamp d. Mesic to wet woods, swamps, meadows. Chiefly E KY and E TN. Frequent. FACW. (*hispida* = stiff hairs). Plate 492.

**R. idaeus** L. subsp. **strigosus** (Michx.) Focke. Gray-leaf red r. High-elevation openings. BR of TN. (*idaeus* = from Mt. Ida in Crete or Turkey; *strigosus* = with short stiff hairs). FACU–. Plates 493 and 494.

**R. longii** Fernald. Long's b. Disturbed areas, often forming large, impenetrable thickets. Occasional in KY and TN. (*longii* = after Bayard Long, one of its discoverers). Indistinguishable from *R. bifrons* in winter, unless old fruiting branches are present; the inflorescence is many-fruited, usually > 10 in *R. bifrons*, and few-fruited, usually fewer than 10 in *R. longii*. FAC. Plates 495 and 496.

**R. occidentalis** L. Black r. Dry to mesic woods, fields, and thickets. Across KY and TN. Frequent. (*occidentalis* = western, occidental, of the West). Plate 497.

**R. odoratus** L. Flowering r. Mesic woods and woodland borders. Chiefly E KY and E TN. Rare. (*odoratus* = fragrant, sweet scented, bearing perfume). Plate 498.

**R. pensilvanicus** Poir. Pennsylvania b. Open woods, pastures, roadsides. Chiefly IP of KY, very rare in IP of TN. Infrequent. Not pictured.

** **R. phoenicolasius** Maxim. Wineberry. Disturbed roadsides and woodlands. Chiefly E KY and E TN. Frequent, naturalized from Asia. Lesser Threat in TN. (*phoenicolasius* = red-purple-haired). Plate 499.

**R. trivialis** Michx. Coastal Plain d. Dry sandy fields and stream banks. Chiefly CP of both states, rare eastward. Infrequent. FACU, FAC. (*trivialis* = common, ordinary, wayside, of crossroads). Plate 500.

**Salix** L. Willow. Family Salicaceae. (*Salix* = the Latin name for willows). Shrubs or trees; terminal buds absent, the smaller branches often breaking and leaving branch scars; lateral buds appressed against twig, with single bud scale; leaf scars alternate, slightly raised, triangular, with 3 bundle scars; stipular scars present but sometimes minute; fruit a capsule with hairy seeds, released in the spring. Remnant leaves are simple, mostly narrowly to broadly lanceolate, pinnately veined, and toothed, or nearly entire in some species. Willows, like poplars, contain the chemical salicin, and can be used for pain (typically as a tea prepared from the bark of twigs). Caution—willows are known to uptake other chemicals in the environment, including toxic metals. The branches and young stems are useful for constructing arrows and the inner bark is a good source of fiber for ropes and string.

Note: A few other species of *Salix*, in addition to those species keyed below, have been documented in the region, including two Historical species in KY—*S. amygdaloides* Andersson and *S. discolor* Muhl., and several exotic species, including *S. fragilis* L., *S. pentandra* L., and *S. purpurea*. These taxa, however, are very rare to possibly extirpated.

1. Bud scale margins free and overlapping, the buds glabrous, sharp-pointed; twigs brittle at base.
    2. Stipules prominent, persistent, or leaving conspicuous scars; shrubs _____ **S. caroliniana.**
    2. Stipules small, deciduous, the scars inconspicuous; trees _____ **S. nigra.**
1. Bud scale margins fused, the buds glabrous or pubescent, mostly blunt; twig flexiblilty various.
    3. Trees; bark furrowed or scaly; twigs and buds glabrous or nearly so; non-native species.
        4. Twigs green, brittle at base; branches often pendulous, "weeping" _____ **S. babylonica.**
        4. Twigs yellowish to dark brown, flexible at base; branches pendulous or spreading _____ **S. alba.**
    3. Shrubs or small trees; bark usually smooth (if furrowed-scaly, then buds pubescent); twigs and/or buds often pubescent; native species except for *S. cinerea*.
        5. Small trees with furrowed or scaly bark at maturity; buds pubescent, to 1 cm long and wider than the twigs; stipules often persistent at least until late fall.
            6. Long ridges absent under bark _____ **S. eriocephala.**
            6. Long ridges visible under the bark _____ **S. cinerea.**
        5. Shrubs with smooth bark; buds much smaller, variously pubescent or glabrous, usually narrower than the twig.
            7. Twigs and buds glabrous or nearly so; twigs yellow-brown; stipular scars inconspicuous _____ **S. exigua.**
            7. Twigs and/or buds notably pubescent; twig color and width various; stipular scars conspicuous.
                8. Twigs and buds with dense gray pubescence; twigs flexible at base; colonial shrub, often < 1 m tall _____ **S. humilis.**
                8. Twigs and buds lacking dense gray pubescence; twigs brittle at base; shrub > 1 m tall _____ **S. sericea.**

\* S. alba L. White w. Moist, disturbed sites. Infrequent. Cultivated, from Eurasia. Across KY and TN. FACW, FACW–. Plate 501.

\* S. babylonica L. Weeping w. Moist, disturbed sites. Cultivated, from China. Infrequent. Across KY and TN. FACW-, FACW. (*babylonica* = from Babylon). Plates 502 and 503.

S. caroliniana Michx. Carolina w. Floodplains, stream banks. Frequent. Across KY and TN. OBL. Plate 504.

\* S. cinerea L. Gray w. Moist, disturbed sites. Naturalized, from Eurasia. Across KY and in E TN. Rare. (*cinerea* = ash-gray). Not pictured.

S. eriocephala Michx. Heartleaf w. Stream banks, low areas. W and N KY and W TN. FACW. (*eriocephala* = woolly-headed). Plate 505.

S. exigua Nutt. Sandbar w. River bars, floodplains, stream banks. Across KY and TN except for far-eastern regions. OBL. (*exigua* = very small). Plate 506.

S. humilis Marshall. Prairie w. Dry to wet barrens-like habitats. Across KY and TN. FACU. (*humilis* = low-growing). Two varieties: var. **humilis** and var. **microphylla** (Andersson) Fernald, distinguished by leaf and pistillate ament lengths. Plate 507.

S. nigra Marshall. Black w. Floodplains, river banks. Across KY and TN. FACW+, OBL. Plate 508.

S. sericea Marshall. Silky w. Stream banks, wet meadows, moist upland sites. Across KY and TN. OBL. (*sericea* = silky-hairy). Plate 509.

**Sambucus** L. Elderberry. Family Caprifoliaceae or Adoxaceae. (*Sambucus* = Greek name for the elder tree). Shrubs, the twigs thick and with conspicuous lenticels and large soft pith; terminal buds absent; lateral buds ovoid, with paired scales, sometimes clustered; leaf scars opposite, broadly triangular and connected at the edges, bundle scars usually 5; stipular scars absent; fruit a black juicy berry in large terminal clusters, maturing in the summer. Athough there is a long history of using the flower clusters for food (battered and deep-fried), as well as using the ripe fruits (deep purple or black) for jelly and wine, caution is urged. The roots, stems, leaves, and unripe fruits are toxic with cyanide-like compounds. A good practice is to only consume the ripe berries after cooking. Children have also been poisoned from playing with peashooters made from the hollowed-out stems.

1. Pith white; buds to 4 mm long, widest at base; fruits blue-black, in flat-topped clusters ___**S. canadensis.**
1. Pith brown; buds to 10 mm long, constricted at base; fruits red, in elongate clusters_____ **S. racemosa.**

S. canadensis L. Common e. Mesic woods and thickets. Across KY and TN. Frequent. FACW–. Plates 510 and 511.

! S. racemosa L. subsp. **pubens** (Michx.) House. Red e. High-elevation forests and deep gorges. Chiefly in BR of TN, also at scattered sites in AP of both states and in SE AP of KY. Endangered in KY. FACU. (*racemosa* = having racemose inflorescences; *pubens* = softly hairy). Plates 512 and 513.

**Sassafras** Nees. Family Lauraceae. (*Sassafras* = from the Spanish name, *salsafras*, for its medicinal use in breaking bladder and kidney stones). Tree, rarely over 15 m tall, but occasionally to 30+ m tall, with mature bark reddish-brown and deeply furrowed, and all parts spicy-aromatic; twigs brittle, greenish or reddish; terminal bud present, 5–10 mm long, ovoid and green, with about 4 imbricate, dark-margined scales, lateral buds much smaller; leaf scars alternate, raised, half-round to nearly round with bundle scar single or sometimes broken; stipular scars absent; fruit a blue drupe to 8 mm long, falling before the winter. Remnant leaves are simple, ovate to obovate, palmately veined, lobed or unlobed, and entire. This species has a long history of use for foods (boiled roots for tea, dried roots and leaves for flavoring), medicinal tonics, and other herbal uses. The twigs make a useful toothbrush (chew the ends until they are frilly). Caution—recent studies indicate that the safrole oil can be carcinogenic.

**S. albidum** (Nutt.) Nees. Sassafras. Open woodlands and roadsides. Across KY and TN. Frequent. FACU–, FACU. (*albidum* = white). Plate 514.

**Schisandra** Michx. Family Schisandraceae. (*schisandra* = divided man). Woody vines climbing by twining, the bark becoming flaky in older stems; twigs red-brown, with tan to brown pith, twigs lined below the leaf scars; terminal buds present, scaly-imbricate, the buds ovoid; leaf scars alternate, rounded, with 3 bundle scars; stipular scars absent; fruit a red berry. Remnant leaves are simple, elliptic, pinnately veined, and entire to sparsely toothed.

! **S. glabra** (Brickell) Rehder. Bay star-vine. Mesic forests, bluffs. AP of KY, CP of TN. Endangered in KY and Threatened in TN. (*glabra* = smooth, without hairs). Plate 515.

**Sibbaldiopsis** Rydb. Family Rosaceae. (*sibbaldiopsis* = similar to *Sibbaldia procumbens*, and in honor of Robert Sibbald, 1643–1720, Professor of Medicine at Edinburgh). Subshrub, mostly under 10 cm tall, forming matlike colonies; leaves evergreen, alternate, trifoliolate, each leaflet 3-toothed at the tip, shiny-green to reddish in the winter; fruit an achene in stalked clusters, persistent.

! **S. tridentata** (Aiton) Rydb. Dwarf fivefingers. High-elevation openings. BR of TN. Special Concern in TN. (*Potentilla tridentata* Aiton). (*tridentata* = 3-pronged). Plate 516.

**Sideroxylon** L. Buckthorn. Family Sapotaceae. (*Sideroxylon* = iron-wood). Shrub or small tree to 15 m tall, the bark scaly to platy, with red-brown inner bark; twigs gray to purplish, with spur shoots or nodal spines above or to side of the leaf scars, the spines often bearing leaves; milky sap present in growing season; terminal buds present, obtuse and smooth; leaf scars alternate, rounded, the bundle scars obscure; stipular scars absent; fruit a blackish drupe to 1.5 cm long. Remnant leaves simple, elliptic to oblanceolate, pinnately veined, and entire or nearly so.

**S. lycioides** L. Southern b. Dry to alluvial woods and thickets, often associated with limestome sites. C and W KY and TN, absent from E KY and rare in E TN. Infrequent. FACW. [*Bumelia lycioides* (L.) Pers.]. (*lycioides* = from Lycia). Plates 517 and 518.

**Smilax** L. Greenbrier, cat brier, saw brier. Family Smilacaceae. (*Smilax* = ancient Greek name for a scrape, referring to the prickly or bristly stems). Woody vines climbing by pairs of tendrils attached to the base of petioles; stems usually prickly, lacking a pith, the vascular bundles scattered; buds 3-angled, with a single scale; leaves alternate, but leaf scars absent, the leaves breaking off above the base of the petiole, which persists as a petiolar stalk; fruit a berry, black or red, in umbel-like clusters, the fruits often persistent into winter. Evergreen or persistent leaves often present, these simple, lobed in some, the margins entire or sometimes prickly or bristly, with arcuate venation (the veins arching toward apex). The stems, leaves, and tendrils are edible, but are better in spring and early summer when they are fresh. The greenbriers were used widely by Native Americans for both food and medicine. The powdered roots were use to make a thickening agent and as flour for making bread; the roots were also used to prepare a variety of herbal treatments (sores and burns, as a tonic for gout, and for stomach troubles). Apparently the ripened berries were not widely used, and there are few reports in the literature of their uses as food. The stems are a good source for cordage.

1. Leaves leathery and evergreen, oblong to narrowly ovate, the bases cuneate to rounded, the midvein more prominent than lateral veins; berries immature in winter, the mature berries black _____ **S. laurifolia.**

1. Leaves firm but not leathery, deciduous, but persistent into winter in some species, the shape more broadly ovate to orbicular-cordate to 3-lobed, the midvein and lateral veins about equally prominent; mature berries often present in winter.
   2. Berries red, on peduncle to 1.5 cm long; stems slender with few prickles, usually sprawling; plants very rare in wetlands of W TN _____ **S. walteri.**
   2. Berries black, the peduncles various; stems conspicuously prickly or bristly, often high climbing; plants generally common throughout in lowlands and uplands.
      3. Leaves usually present and white-glaucous below; stems white-waxy and rounded in cross-section; peduncles much longer than petioles; berries with white-waxy covering _____**S. glauca.**
      3. Leaves usually absent, or if present, then lacking white-glaucous coating, typically light green below; stems green and rounded or angled; peduncles and berries various, but berries lacking white-waxy covering.
         4. Peduncles to 1.5 cm long, equaling or slightly longer than the petioles; stems strongly prickly, but lacking needlelike bristles; leaf margins entire or finely serrulate, not spiny-margined _____**S. rotundifolia.**
         4. Peduncles > 1.5 cm long, notably longer than the petioles; plants either with bristly stems or spiny leaf margins.
            5. Stems with numerous needlelike, often blackish bristles, especially on lower parts; leaf margin entire or serrulate, lacking a thickened band along the margin; stems rounded to many-angled; berry 1- or 2-seeded _____**S. tamnoides.**
            5. Stems lacking needlelike bristles; leaf margins often bristly or spiny, with a thickened band along the margin; stems distinctly 4-angled; berries 1-seeded_____**S. bona-nox.**

**S. bona-nox** L. Saw g. Dry woods and thickets. Across KY and TN. FACU, FAC. (*bona-nox* = good night). Plates 519 and 520.

**S. glauca** Walter. Cat g. Dry to mesic woods and thickets. Across KY and TN. FACU. (*glauca* = whitish bloom). Plates 521 and 522.

! **S. laurifolia** L. Laurel g. Lowlands. Known only from 2 counties in SE TN. FACW. Special Concern in TN. (*laurifolia* = laurel-leaved). Plates 523 and 524.

**S. rotundifolia** L. Common roundleaf g. Dry to mesic woods and thickets. Across KY and TN. FAC. (*rotundifolia* = round-leaved). Plates 525, 526, and 527.

**S. tamnoides** L. Bristly g. Dry to mesic woods and thickets. Across KY and TN. FAC. (*S. hispida* L.). (*tamnoides* = tamnus-like, Latin referring to climbing plants). Plate 528.

**S. walteri** Pursh. Red-berried g. Lowlands of CP. Known only from one county in W TN. OBL. (for Thomas Walter, 1740–1789, of South Carolina). Plate 529.

**Solanum** L. Family Solanaceae. (*Solanum* = soothing or comforting). Woody vines (worldwide the genus includes herbs, shrubs, and trees) climbing by scrambling or twining; stems 5-angled, hollow (or pith spongy), and inner bark strongly malodorous; buds rounded, white-hairy, leaf scars alternate, circular, with single bundle scar; stipular scars absent; fruit a red berry in drooping clusters. Both foliage and fruit are poisonous.

** **S. dulcamara** L. Climbing nightshade. Disturbed places, thickets. Escaped at few sites in N KY and in C and E TN. Naturalized from Eurasia. Lesser Threat in KY. FAC–. (*dulcamara* = bitter-sweet). Plates 530 and 531.

**Sorbus** L. Family Rosaceae. (*Sorbus* = the ancient Latin name for the fruit of the service tree). Small tree to 12 m tall, with smooth gray-brown bark; the twigs stout with prominent lenticels, brown pith, and cherry-like odor; terminal buds present, to 18 mm long, conical, gummy and glossy, purple or red, with 3 or 4 imbricate scales, the lateral buds much smaller; leaf scars alternate, raised, narrowly linear, with 5 bundle scars; stipular scars absent; fruit a red pome, to 8 mm thick, in clusters persisting into winter. The fruits can be eaten raw, but are better after exposure to a couple of freezing periods; they can also be stewed and made into a sauce. Remnant leaves may be present; these are pinnately compound and sharply toothed.

**S. americana** Marshall. American mountain-ash. Open sites at high elevations of BR in TN. FACU. Plates 532 and 533.

**Spiraea** L. Spiraea. Family Rosaceae. (*Spiraea* = garland). Shrubs to 3 m tall, the stems often stiff and few-branched; terminal bud usually absent, the twigs often terminating in fruit clusters; buds small, globose, about 6-scaled, the pubescence concealing the scales in some species, collateral buds sometimes present; leaf scars alternate, raised, crescent to half-round, with single bundle scar; stipular scars absent; fruit a small follicle about 3 mm long. Remnant leaves simple, elliptic to lanceolate, pinnately veined, and toothed.

1. Dried fruits/flowers arranged along main stem in unbranched flat-topped clusters _____**S. prunifolia.**
1. Dried fruits/flowers arranged in branched clusters at apex of stem.
    2. Inflorescence longer than broad.
        3. Stems and follicles brown-woolly _____**S. tomentosa.**
        3. Stems and follicles glabrous or only lightly pubescent _____**S. alba.**
    2. Inflorescence wider than long, often flat-topped.
        4. Buds blunt; leaves blunt, nearly entire _____**S. virginiana.**
        4. Buds acute; leaves acuminate, doubly-serrate_____**S. japonica.**

! **S. alba** Du Roi. White meadowsweet. Swamps, wet openings, and shores. Known only from 3 AP counties (2 in TN and 1 in KY) and from 1 county in W KY. Historical in KY and Endangered in TN. FACW+. Plates 534 and 535.

** **S. japonica** L.f. Japanese s. Mesic or wet woods. Chiefly AP of KY and TN, scattered in IP of both states. Infrequent, naturalized from Japan. Significant Threat in KY and TN. FACU–, FACU+. Plates 536 and 537.

* **S. prunifolia** Siebold & Zucc. Plumleaf s. Disturbed areas, old home sites, occasionally in woodlands. Across TN and to be expected in KY. Infrequent, introduced from Asia. (*prunifolia* = plum-leaved). Plate 538.

**S. tomentosa** L. Hardhack. Swamps, marshes, and meadows. Chiefly AP of both states, scattered across KY and across C and E TN. Infrequent. FACW. (*tomentosa* = thickly matted with hairs). Plates 539 and 540.

! **S. virginiana** Britton. Appalachian s. Along rocky stream banks. AP of KY and TN. Threatened in KY, Endangered in TN, Threatened in U.S. FACU, FACW. Plate 541.

**Staphylea** L. Family Staphyleaceae. (*Staphylea* = cluster). Shrub to 5 m tall, the bark greenish or grayish and white-striped; terminal buds absent, the twigs usually tipped by a pair of buds, these ovoid, about 5 mm long and 4-scaled; leaf scars opposite, not connected by lines, crescent to triangular, with 3 bundle scars; stipular scars present; fruit a bladdery 3-lobed capsule, to 5 cm long, with loose seeds that rattle when shaken.

**S. trifolia** L. Bladder-nut. Mesic woodlands. Across KY and TN. Frequent. FAC. (*trifolia* = with three leaves, in reference to the three leaflets). Plates 542 and 543.

**Stewartia** L. Stewartia. Family Theaceae. (*Stewartia* = for John Stewart, 1713–1792, Third Earl of Bute and patron of botany). Shrub or small tree to 7 m tall, the bark shreddy; twigs with spongy (porous) pith; terminal buds present, to 1 cm long, flattened and twisted toward apex, the 2 bud scales covered by silvery-silky hairs; leaf scars alternate, in one plane, half-round, with single large bundle scar; stipular scars absent; fruit a 5-chambered capsule to 2 cm long. Remnant leaves simple, ovate to elliptic, pinnately veined, and entire to slightly toothed.

**S. ovata** (Cav.) Weatherby. Mountain camellia. Dry to mesic forests. SE KY and E TN. Rare. (*ovatus* = egg-shaped). Plate 544.

**Styrax** L. Snowbell. Family Styracaceae. (*Styrax* = ancient Greek name for storax gum). Shrubs or small trees to 7 m tall, the bark smooth and gray when young, fissured in older individuals; twigs stellate-pubescent; terminal buds absent, appearing naked or 2-scaled, the lateral buds often superposed; leaf scars alternate, in one plane, with single bundle scar; stipular scars absent; fruit dry and drupelike, to 1 cm long, splitting to reveal reddish seeds in winter. Remnant leaves simple, elliptic to obovate, pinnately veined, and entire or slightly toothed.

1. Twigs stout, to 3 mm thick; fruit to 10 mm thick _____ **S. grandifolius.**
1. Twigs slender, to 1.5 mm thick; fruit to 8 mm thick _____ **S. americanus.**

**S. americanus** Lam. American s. Swamps and wet woods. CP and W IP of KY and TN. Infrequent. OBL, FACW. Plate 545.

! **S. grandifolius** Aiton. Bigleaf s. Mesic to dry woodlands. Across C and E TN, but known only from single county in W IP of KY. FACU, FACU–. Endangered in KY. (*grandifolius* = with large leaves). Plate 546.

☠ **Symphoricarpos** Duhamel. Family Caprifoliaceae. (*Symphoricarpos* = clustered fruits). Shrubs to 2 m tall with very slender branches, the bark shreddy when older; twigs pubescent, with bud scales persistent at twig bases, the pith sometimes hollow; terminal bud absent; lateral buds to 3 mm long, sometimes clustered, rounded, the scales paired and keeled; leaf scars opposite, raised, connected by lines, half-round or torn, with bundle scar single but obscure; stipular scars absent; fruit a red or white berry, in axillary clusters, and persistent into winter, but ingestion may result in digestive problems. Remnant leaves simple, ovate to elliptic, pinnately veined, and entire to slightly lobed.

1. Berries white; pith hollow _____ **S. albus.**
1. Berries red; pith continuous _____ **S. orbiculatus.**

! **S. albus** (L.) S.F.Blake. Snowberry. Open, rocky limestone woods of C KY. Endangered in KY. FACU–. Not pictured.

**S. orbiculatus** Moench. Coralberry. Woodlands and fields, chiefly in limestone areas. Across both states, but chiefly in C KY and TN. Frequent. UPL, FAC–. (*orbiculatus* = disc-shaped, circular in outline). Plates 547 and 548.

**Symplocos** Jacq. Family Symplocaceae. (*Symplocos* = united). Shrub or small tree, the bark gray and warty; twigs pubescent, angled, with chambered pith; terminal buds present, to 5 mm long, with keeled scales, the buds pubescent and clustered at twig tips; lateral buds of two kinds, pointed leaf buds and rounded flower buds; leaf scars alternate, half-round, with single bundle scar; stipular scars absent; fruit a green drupe to 12 mm long. Leaves, often persistent or semi-evergreen, are simple, elliptic to obovate, pinnately veined, and wavy-margined.

! **S. tinctoria** (L.) L'Hér. Common sweetleaf. Wet woods and stream margins. Few sites in W and E TN, to be expected in S KY. Special Concern in TN. FAC. (*tinctoria* = used for dyeing). Plates 549, 550, and 551.

**Syringa** L. Family Oleaceae. (*Syringa* = pipe). Shrub to 7 m tall, with rough bark; terminal buds absent, the twigs often tipped by a pair of buds, these to 8 mm long, the scales thick, in about 4 pairs, greenish to purplish and keeled; leaf scars opposite, not connected by lines, with single bundle scar; stipular scars absent; fruit a capsule to 15 mm long, persistent in winter. Remnant leaves simple, ovate with truncate or cordate base, pinnately veined, and entire.

\* **S. vulgaris** L. Common lilac. Cultivated from Europe, persisting at old homesites and sporadically escaping at widely scattered sites in both states. (*vulgaris* = usual, common). Plate 552.

**Taxodium** Rich. Family Cupressaceae. (*Taxodium* = similar to *Taxus,* yewlike). Tree to 40 m tall, with shreddy bark, the base of trunk often swollen and buttressed, the roots producing the characteristic "knees" protruding above water in swampy habitats; twigs slender, with tiny globose buds; branch scars present and alternate, the leaf scars absent, but stiff pointy scales often present and associated with clusters of knobby buds and nodes; seed cones spherical, to 2.5 cm broad; immature pollen cones (these with pinelike odor) in pendulous panicles, and present in winter.

**T. distichum** (L.) Rich. Bald-cypress. Swamp forests. Chiefly CP of both states, also in W IP and scattered eastward at wet sites. Infrequent. OBL. (*distichum* = in two alternately opposed ranks). Plates 553 and 554.

**Taxus** L. Family Taxaceae. (*Taxus* = the ancient Latin name for yew from Dioscorides). Evergreen shrubs or small trees; twigs green; needles alternate, flat, sharp-tipped, about 30 mm long by 2 mm wide, the bases decurrent on the twig; scaly cones absent, the seed partly enclosed by a fleshy red aril, about 6 mm wide. Leaves, seeds, and all parts highly poisonous except fleshy red aril around the seed. The foliage was used (in tiny amounts) to treat various ailments by the Native Americans. In recent years the genus has been investigated as a source of chemicals to treat ovarian cancer. The yew is among the most prized for making high-quality bows.

! **T. canadensis** Marshall. Canadian yew. Cold air drainages, stream sides, and on bluffs, in mixed mesophytic forests. AP. Threatened in KY (7 counties) and Endangered in TN (1 county). FAC (in KY). This species is low-growing, mostly to 2 m tall, and is atypical of the genus in that it is monoecious (producing male and female reproductive structures) on the same individual. Non-native species of *Taxus*, including *T. baccata* L. and *T. cuspidata* Siebold & Zucc., are often cultivated but rarely escape. These non-native species can be taller and more treelike, and are dioecious (plants unisexual). Plates 555 and 556.

**Thuja** L. Arbor-vitae, white-cedar. Family Cupressaceae. (*Thuja* = Theophrastus' name for a resinous, fragrant-wooded tree). Evergreen tree mostly to about 20 m tall, sometimes > 30 m tall, with fibrous and fissured bark; branchlets flattened; leaves opposite, scale-like, 2–3 mm long, 4-ranked, appressed and overlapping, the ones in lateral rows keeled, and the ones in dorsal and ventral rows flattened; cones oblong, about 1 cm long, with brownish dry scales in about 4 pairs. The foliage and bark were widely used by Native Americans for teas and other concoctions, but in high concentrations these brews can cause convulsions or even death.

! **T. occidentalis** L. Northern w.-c., arborvitae. Stream banks and bluffs. S AP of KY and across northern tier of E TN counties. Threatened in KY and Special Concern in TN. FACW, UPL. (*occidentalis* = western, of the West). Plate 557.

**Tilia** L. Basswood. Family Tiliaceae or Malvaceae. (*Tilia* = wing). Tree to 40 m tall, with furrowed bark, and characteristic sprouting from trunk base; twigs zigzag, greenish or reddish; terminal buds absent; lateral buds ovoid, greenish or reddish, with 2 or 3 visible scales; leaf scars alternate, in one plane, half-round, with bundle scars > 3; stipular scars conspicuous and unequal; fruit hard and nutlike, pea-sized, borne in a cluster on a peduncle attached to a leafy bract. Remnant leaves are simple, ovate to orbicular, cordate-oblique at base, pinnately veined, and toothed (the lateral veins run straight to the teeth, unlike the similar-leaved *Morus* spp., in which the lateral veins loop before reaching the teeth). The winter buds, inner bark, and young leaves can be eaten, and the bark is a good source of fiber for making ropes and string.

**T. americana** L. Basswood. Mesic woods. The two varieties: var. **americana**, American b., and var. **heterophylla** (Vent.) Loudon, white b., are distinguishable in winter by their withered leaves, with the latter variety having lower leaf surfaces distinctly whitened-pubescent. Both varieties occur across both states, with the var. *heterophylla*, in particular, less frequent westward. FACU. (*heterophylla* = different-leaved, referring to the contasting whiter lower leaf surface). Plate 558.

**Toxicodendron** Mill. Family Anacardiaceae. (*Toxicodendron* = poison tree). Woody vines, shrubs, or small trees; terminal bud present, naked or with valvate scales, larger than lateral buds; leaf scars alternate, crescent- or shield-shaped, with > 3 bundle scars; stipular scars absent; fruit a white drupe in axillary clusters, persistent into winter. All parts contain an oily resin, and contact can result in rash, blisters, or more severe reactions. The twigs and foliage should not be burned, even in winter, because the smoke can carry droplets of the resin and stimulate a severe response in susceptible individuals. Cutting firewood with entangled poison-ivy can also lead to exposure.

1. Small trees to 7 m tall; buds glabrous and scaly _____ **T. vernix.**
1. Shrubs or vines; buds hairy.
    2. Shrubs, usually < 1 m tall; the stems lacking aerial roots.
        3. Twigs glabrous; fruits glabrous or slightly pubescent _____ **T. radicans.**
        3. Twigs and fruits distinctly pubescent _____ **T. pubescens.**
    2. Vines, climbing by aerial roots _____ **T. radicans.**

**T. pubescens** Mill. Poison-oak. Documented from dry woods across S TN. FACU. (*pubescens* = hairy). Plates 559 and 560.

**T. radicans** (L.) Kuntze. Poison-ivy. Upland and lowland woods and thickets. Across KY and TN. Abundant. FAC. (*Rhus radicans* L.). (*radicans* = with rooting stems or leaves). Plates 561 and 562.

! **T. vernix** (L.) Kuntze. Poison-sumac. Swamps and wet woods. AP. Infrequent in TN, rare in KY. Endangered in KY. OBL. (*Rhus vernix* L.). (*vernix* = varnish, Old French, *vernis*, from medieval Latin *vernix* for fragrant resin). Plates 563, 564, and 565.

**Trachelospermum** Lem. Family Apocynaceae. (*Trachelospermum* = necked seed). Woody vine with milky sap and smooth bark; twigs slender and reddish; terminal buds absent; lateral buds small and acute, dark and scurfy; leaf scars opposite, raised and connected by a line, half-round, with single bundle scar; stipular scars absent; fruit a pair of elongate follicles to 20 cm long, with hairy seeds. Remnant leaves simple, lanceolate to obovate, pinnately veined, and entire.

**T. difforme** (Walter) A.Gray. Climbing dogbane. Swamps, wet woods, along rocky stream banks. Chiefly CP and W IP of both states. Rare. FACW. (*difforme* = of unusual or abnormal form or shape). Plates 566 and 567.

⚕/☘ **Tsuga** Carrière. Hemlock. Family Pinaceae. (*Tsuga* = from the Japanese vernacular name for the hemlock cedar). Evergreen tree to 35 m tall, the older trunks with bark fissured and with rounded ridges; needles alternate, flat, blunt, with 2 white stomatal bands below, to 13 mm long, with slender petiole attached to a raised peg-like projection on twig; cones scaly, with blunt edges. Both of these species are now severely threatened by hemlock woolly adelgid. Native Americans boiled the fresh leaves, twig tips, and inner bark for making tea, which was considered helpful for respiratory ailments.

1. Needles entire, spreading in several directions; cones 4 cm long _____**T. carolinina.**
1. Needles minutely toothed near tips, spreading in 1 plane; cones to 2.5 cm long_____**T. canadensis.**

**T. canadensis** (L.) Carrière Eastern h. Mixed mesophytic forests. E KY and E TN, scattered at disjunct sites westward in IP of KY. Frequent. FACU. Plates 568 and 569 (showing hemlock woolly adelgid).

! **T. caroliniana** Engelm. Carolina h. Woods and bluffs. N BR of TN. Threatened in TN. Plate 570.

⚕/☘/➡ **Ulmus** L Elm. Family Ulmaceae. (*Ulmus* = the ancient Latin name for elms). Trees; twigs zigzag, sometimes corky-winged; terminal buds absent; lateral buds offset to one side of leaf scar, with about 6 imbricate-scaly buds; leaf scars alternate, in one plane, crescent or half-round, with 3 sunken bundle scars, or these sometimes broken into > 3; stipular scars present and unequal; fruit a rounded samara. Remnant leaves simple, ovate to elliptic, with oblique base, pinnately straight-veined, and doubly-toothed. The mucilaginous inner bark of red (slippery) elm has a long history of use, both internally (for respiratory and gastrointestinal problems) and externally (for wound healing). The inner bark can also be eaten in emergency situations, and is a good source of fiber for cordage. The branches and young stems are also useful for constructing bows and arrows.

1. Twigs, at least some (note older twigs and branches), with corky wings; flowering and fruiting periods various.
    2. Buds to 8 mm long; plants chiefly of upland limestone regions_____ **U. thomasii.**
    2. Buds to 6 mm long; plants of various distributions.
        3. Bundle scars 3; plants of southern and western distributions.
            4. Samaras absent in the fall; widespread distribution_____**U. alata.**
            4. Samaras present in the fall; restricted distribution in W TN_____ **U. crassifolia.**
        3. Bundle scars 4–6; buds glabrous; plants of central distribution, the populations widely scattered, rare and local _____**U. serotina.**
1. Twigs lacking corky wings on younger and older twigs and branches; flowering and fruiting in the spring.
    5. Twigs gray and pubescent; mature bark uniform in color; buds obtuse, red-tomentose_____ **U. rubra.**
    5. Twigs brownish or reddish, glabrous or nearly so; mature bark with light and dark layers, candy-stripe in appearance; buds acute, smooth or slightly pubescent_____**U. americana.**

**U. alata** Michx. Winged e. Dry to mesic woods, along streams. Across KY and TN. Frequent. FACU, FACU+. (*alata* = wing-like). Plates 571 and 572.

**U. americana** L. American e. Mesic to wet woodlands. Across KY and TN. Frequent. FACW–, FACW. Plate 573.

! **U. crassifolia** Nutt. Cedar e. Mesic woods and bluffs. Known only from 2 counties in W TN. Special Concern in TN. (*crassifolia* = thick, fleshy). Plate 574.

**U. rubra** Muhl. Red e., slippery e. Mesic woodlands of slopes and bottoms. Across KY and TN. Frequent. FAC. Plate 575.

! **U. serotina** Sarg. September e. Dry to mesic woodlands of limestone bluffs, slopes, and along streams. Across S KY and IP of TN. Special Concern in KY. FAC+, UPL. (*serotina* = autumnal). Plate 576.

**U. thomasii** Sarg. Rock e. Dry to mesic woodlands of rocky limestone slopes and lowlands. Chiefly BG of KY and CB of TN, scattered elsewhere. Infrequent. FACU+, FAC. (*thomasii* = for David Thomas, American civil engineer in early nineteenth century). Plates 577 and 578.

🌿 **Vaccinium** L. Blueberry, cranberry. Family Ericaceae. (*Vaccinium* = a Latin name of great antiquity with no clear meaning). Shrubs, varying from low and trailing to small trees; terminal buds absent; buds scaly, either valvate or imbricate; leaf scars alternate, half-round, with single bundle scar; stipules absent; fruit a fleshy berry with calyx remnant at top of fruit. Remnant leaves are simple, lanceolate to elliptic to obovate, pinnately veined, and entire or toothed. The fruits are edible but vary in taste—the worst are those of *V. arboreum* and *V. stamineum*, and the best are those of the true blueberries (*V. corymbosum* and allies, and *V. pallidum*). *V. macrocarpon* is very tasty but highly restricted in distribution, whereas highbush cranberry (*V. erythrocarpum*) can be abundant at high elevations in the southern Appalachians and also has a palatable berry. Blueberries continue to be touted as one our most healthy food items; they are very high in antioxidants, and appear to help circulation throughout the body and to prevent clogging of the arteries.

1. Plants prostrate, with aerial branches < 20 cm long; leaves evergreen, < 2 cm long and 1 cm wide; berries red _____ **V . macrocarpon.**
1. Plants with erect or ascending branches > 20 cm tall; leaves deciduous, or if persistent, then plants much taller; berry color various, mostly black or blue.
    2. Plants < 1 m tall (lowbush blueberries with warty-dotted twigs).
        3. Twigs with long, straight trichomes _____ **V. hirsutum.**
        3. Twigs glabrous and green _____**V. pallidum.**
    2. Plants 1–10 m tall; twig surfaces various.
        4. Small trees, with well-developed trunk, often 5–10 m tall; bark reddish; leaves often persistent, shiny and leathery, to 6 cm long_____**V. arboreum.**
        4. Shrubs with many stems, usually much less than 5 m tall; bark various; leaves deciduous [except sometimes in small-leaved (< 3 cm long) *V. elliottii*].
            5. Shrubs to 2 m tall; buds lanceolate, about 5 mm long, the scales 2 and valvate, gradually tapered to the tip; twigs and buds smooth; berries dark red, solitary in leaf axils; high-elevation species _____ **V. erythrocarpum.**
            5. Shrubs to 5 m tall; buds blunt or acute, 2–3 mm long, and often with abruptly pointed scale tips, the scales 3 or more and imbricate, twig pubescence various, often hairy; berries of various colors, in racemes; mostly widespread species.
                6. Buds about as wide as long; twig surfaces lacking warty dots; berries yellow, to pink to black, in bracteate racemes_____ **V. stamineum.**
                6. Buds longer than wide, with sharp-tipped scales; twig surfaces with warty dots; berries black, in bractless racemes.
                    7. Remnant leaves shiny, to 3.5 cm long by 1.5 cm wide, with margins serrulate, the teeth spine- or gland-tipped_____ **V. elliottii.**
                    7. Remnant leaves dull, usually over 3.5 cm long and 1.5 cm wide, the margins usually entire.
                        8. Twigs glabrous_____**V. corymbosum.**
                        8. Twigs pubescent _____ **V. fuscatum.**

**V. arboreum** Marshall. Sparkleberry. Rocky upland acidic woodlands. Chiefly IP (except BG) in KY, and across TN. Frequent. FACU. (*arboreum* = tree like, branched). Plate 579.

**V. corymbosum** L. High-bush b. Dry upland woods to swamp forests. Chiefly E KY and E TN, scattered westward. Infrequent. FACW–, FACW. Source of many commercial forms of blueberries. (*corymbosum* = corymb-like). Plates 580, 581, and 582.

! **V. elliottii** Chapm. Mayberry. Known only from pond margins in Coffee County, TN. Endangered in TN. FAC+. (after Stephen Elliott, 1771–1830). Plate 583.

! **V. erythrocarpum** Michx. Bearberry. Mountain forests. SE AP of KY, AP and BR of TN. Endangered in KY. FAC. (*erythrocarpum* = with red fruits). Plate 584.

**V. fuscatum** Ait. Black high-bush b. Chiefly acid woods, lowlands to uplands. Scattered across TN. FAC+ . (*fuscatum* = dusky brown). Plate 585.

**V. hirsutum** Buckley. Hairy b. Dry oak-pine woods. S AP and BR of TN. (*hirsutum* = soft-hairy). Plate 586.

! **V. macrocarpon** Aiton. Cranberry. Bogs and open damp areas. Known only from 3 counties in TN . Threatened in TN. OBL. (*macrocarpon* = large-fruited). Plates 587 and 588.

**V. pallidum** Aiton. Low-bush b. Dry acidic woodlands. All areas east of Tennessee River in KY and TN (except CB and BG). Frequent. (*pallidum* = pale). Plate 589.

**V. stamineum** L. Deerberry. Dry to mesic woodlands. Across KY and TN, far western areas, CB, and BG. Frequent. FACU–, FACU. Two varieties: var. **stamineum** has nearly glabrous leaves and twigs; and var. **melanocarpum** C.Mohr, with distinctly pubescent leaves and twigs. (*stamineum* = with prominent stamens). Plate 590.

**Viburnum L.** Viburnum, haw. Family Caprifoliaceae or Adoxaceae. (*Viburnum* = the Latin name for the wayfaring tree). Shrubs or small trees; terminal buds present, naked or scaly, larger flower buds commonly present; leaf scars opposite, meeting or connected by lines, crescent- to V-shaped, with 3 bundle scars; stipular scars absent; fruit a red to blackish-blue drupe, often present in winter. The fruits are variable in taste—some may be bitter or with bad odor. Remnant leaves are simple, of various shapes, pinnately veined except in *V. acerifolium*, and mostly toothed (entire in few species). With their large seeds the fruits do not provide much fleshy food—boiling, straining, dilution, and sweetening can produce a palatable juice or sauce. Arrow shafts can be constructed from the branches and young stems.

1. Leaves evergreen, veiny; buds leafy and naked; non-native species _____ **V. rhytidophyllum.**
1. Leaves deciduous; buds various; native species.
    2. Buds naked (veiny and leaf-like), 1–2 cm long; remnant leaves rounded and cordate, often 20 cm long and wide_____**V. lantanoides.**
    2. Buds scaly; leaves otherwise.
        3. Buds scales 2 (valvate); remnant leaves with veins not reaching leaf margins.
            4. Bark thick and blocky on main branches, these with many short, stubby side branches spreading almost at right angles; buds rusty red or gray to brown with waxy or scurfy surface, to 1.5 cm long; remnant leaves finely toothed.
                5. Buds gray to brown, glabrous or scurfy _____**V. prunifolium.**
                5. Buds with dark-reddish pubescence _____ **V. rufidulum.**
            4. Bark thin and smooth on main branches, these with side branches few, flexuous and ascending, lacking stubby side branches; buds light brown or tannish, with scurfy or matted-hairy surface, to 2 cm long; remnant leaves entire to irregularly serrate.

6. Lateral buds to 6 mm long; plants often rhizomatous, forming spreading colonies; lowland species _____ **V. nudum.**

6. Lateral buds to 10 mm long; plants usually in clumps, not colony-forming; upland species, including at high elevations in BR _____ **V. cassinoides.**

3. Buds with 4 or more scales; remnant leaves with veins extending to marginal tooth.

7. Plants usually < 1 m tall; buds with lowest scales very short; remnant leaves 3-lobed _____
_____ **V. acerifolium.**

7. Plants usually > 1 m tall; buds with lowest scales at least half as long as bud; remnant leaves unlobed.

8. Bark exfoliating; plants restricted to C KY and TN_____**V. molle.**
8. Bark tight; plants with distributions otherwise.

9. Twigs stellate pubescent; widespread plants of uplands and lowlands _____**V. dentatum.**
9. Twigs lacking stellate pubescence; plants of uplands, rare and local in range.

10. Twigs pubescent; terminal buds often constricted at base; remnant leaves with < 12 teeth per side_____ **V. rafinesquianum.**

10. Twigs glabrous; terminal buds usually not constricted; remnant leaves usually with 12 or more teeth per side_____ **V. bracteatum.**

**V. acerifolium** L. Mapleleaf v. Deciduous woodlands. Eastern half of KY and TN. Frequent. UPL, FACU. (*acerifolium* = maple leaved). Plate 591.

! **V. bracteatum** Rehd. Limerock arrow-wood. Rich limestone woods. Known only from Franklin County, TN. Endangered in TN. (*bracteatum* = with bracts). Plate 592.

**V. cassinoides** L. Withe-rod. Stream banks in rich woods. E IP and AP of KY and TN, BR of TN. Infrequent. FACW. [*V. nudum* L. var. *cassinoides* (L.) Torr. & A.Gray]. (*cassinoides* = cherry-like). Plate 593.

**V. dentatum** L. Arrow-wood. Mesic woods, swamps. Chiefly E KY and E TN, scattered westward across KY and on E IP of TN. Infrequent. FAC. (*dentatum* = having teeth, with outward pointing teeth). Plate 594.

! **V. lantanoides** Michx. Witch hobble. Mesic woods or high-elevation forests. SE AP of KY, BR of TN. Endangered in KY. FAC. (*V. alnifolium* Marshall). (*lantanoides* = from *lantana*, an old Latin name for a viburnum). Plate 595.

! **V. molle** Michx. Kentucky v. Calcareous woods, bluffs, and ledges above streams. Most populations in BG of KY, few southward into E IP of TN. Special Concern in KY and Endangered in TN. (*molle* = softly hairy). Plates 596 and 597.

! **V. nudum** L. Possum h. Low mesic woods, stream banks, swamps. Chiefly on AP of TN, also westward to N CP and adjacent county in KY. Endangered in KY. OBL, FACW+. (*nudum* = bare, naked). Plate 598.

**V. prunifolium** L. Plumleaf v. Mesic woods, thickets. Across KY and E TN, much less common westward in TN. Frequent. FACU. (*prunifolium* = plum leaved). Plates 599 and 600.

! **V. rafinesquianum** Schult. Rafinesque's v. Calcareous woods and ledges above streams. Most common in C KY, rare in E KY and E TN. Two varieties: var. **affine** (Bush ex. C.K.Schneid.) House and var. **rafinesquianum**, are indistinguishable in winter (this variety has more pubescent leaves and petioles shorter than stipules), the latter variety being listed as Threatened in KY. (for Constantine S. Rafinesque, 1783–1840, prodigious naturalist who named over 6000 species new to science, mostly vascular plants). Plate 601.

\* **V. rhytidophyllum** Hemsley. Leatherleaf v. Cultivated and rarely escaping to disturbed sites in KY and TN. Native of China. Plate 602.

**V. rufidulum** Raf. Rusty black h. Woodlands and thickets. Across TN and across KY except for far E KY. Frequent. UPL, FACU. (*rufidulum* = somewhat rusty red). Plates 603 and 604.

☙ **Vinca** L. Periwinkle. Family Apocynaceae. (*Vinca* = bond). Evergreen, slightly woody, vinelike shrubs, trailing but not climbing, forming a ground cover, the erect leafy branches < 20 cm long, with milky sap (at least during the growing season); leaves opposite, simple, elliptic to ovate, to 4 cm (–6 cm) long, entire; fruit of follicles but rarely fruiting in the region. The milky sap likely has some toxic effects, but species in the genus are also being investigated for possible anti-cancer properties.

1. Leaves and calyx ciliate; leaves deltoid _____**V. major.**
1. Leaves and calyx entire; leaves lanceolate _____**V. minor.**

** **V. major** L. Greater p. Lawns and old home sites. Escaping across TN and occasional in KY, introduced from Europe. Significant Threat in TN. (*major* = larger, greater, bigger). Plate 605.

** **V. minor** L. Common p. Lawns, old home sites, cemetaries. Escaping across KY and TN, introduced from Europe. Significant Threat in KY and TN. (*minor* = smaller). Plate 606.

☙/☤/➨ **Vitis** L. Grape. Family Vitaceae. (*Vitis* = the Latin name for the grapevine). Woody vines, climbing by tendrils or trailing, most species with flaky or shreddy bark; twigs often angled, with tendrils attached opposite the leaf scars, the pith brownish and diaphragmed at the nodes in most species; terminal bud absent; lateral buds rounded, with 2–4 scales, partly sunken; leaf scars alternate, crescent to half-round, with bundle scars several or indistinct; stipular scars narrow and distinct; fruit a bluish juicy edible berry, with some withered fruits often persistent into winter. Remnant leaves are simple, orbicular to ovate in outline, palmately veined, lobed or unlobed, and toothed. The uncooked fruits may be highly acidic, and should not be eaten in quantity. Note: grapes are many seeded, whereas the poisonous look-alike moonseed (*Menispermum canadense*) fruits have a single crescent-shaped seed. The compounds found in grapes, grape seeds, and wine are considered beneficial to blood flow, heart function, and mental function (including possible protection against Alzheimer's disease). The inner bark is also a good source of fiber for making rope and string. Most species are common across both states, producing sturdy ropelike vines that have often been used by children to swing Tarzanlike through the forest.

1. Bark smooth; pith continuous through the nodes; tendrils unbranched _____**V. rotundifolia.**
1. Bark shreddy; pith interrupted at nodes by a woody partition; tendrils branched.
    2. Tendrils present at 3 or more consecutive nodes _____**V. labrusca.**
    2. Tendrils lacking at each third node, or tendrils absent.
        3. Plants trailing, rarely climbing; tendrils few or absent _____**V. rupestris.**
        3. Plants climbing; tendrils present.
            4. Older stems angled; nodal diaphragms 1.5–3.5 mm thick; nodes often banded with red pigment
            _____**V. cinerea.**
            4. Older stems rounded; nodal diaphragms otherwise; nodes usually lacking red pigment.
                5. Nodal diaphragms < 1.5 mm thick_____**V. riparia.**
                5. Nodal diaphragms 2–6 mm thick.
                    6. Stems purplish or reddish _____**V. palmata.**
                    6. Stems brown or gray _____**V. aestivalis/vulpina.**

**V. aestivalis** Michx. Summer g. Dry to mesic woods, thickets, and roadsides. Across KY and TN. Frequent. FACU, FAC–. (*aestivalis* = of summer). Plates 607 and 608.

**V. cinerea** (Engelm. in A.Gray) Engelm. ex Millardet. Downy or graybark g. Wet woods, thickets, and swamps. Across KY and TN. FACW, FAC+. Two varieties: var. **baileyana** (Munson) Comeaux and var. **cinerea**, are indistinguishable in winter. (*cinerea* = ash-gray). Plates 609, 610, and 611.

! **V. labrusca** L. Fox g. Mesic to wet thickets and woodland borders. Scattered records across KY and eastern half of TN. Threatened in KY. FACU, FAC+. This species is the ancestor of the Concord grape and other cultivars and hybrids (all referred to as *V. labruscana* L.H.Bailey), and these may occasionally escape to disturbed areas. (*labrusca* = a wild vine). This species produces grapes particularly good in size and flavor. Plate 612.

**V. palmata** Vahl. Red or Cat g. Swamps. Chiefly CP and W IP of both states, scattered eastward. Infrequent. FACW– (in TN). (*palmata* = with five or more veins arising from one point, hand-shaped). Plate 613.

**V. riparia** Michx. Riverside g. Wet woods, often along streams. CP, scattered eastward to IP of TN and AP of KY. Infrequent. FACW. (*riparia* = of the banks of streams and rivers). Plates 614 and 615.

**V. rotundifolia** Michx. Muscadine g. Woods, especially dry open woodlands. Across KY and TN. Infrequent. FAC–, FAC. (*rotundifolia* = having rounded leaves). Plate 616.

! **V. rupestris** Scheele. Sand g. Rocky habitats along streams. AP, IP. Threatened in KY and Endangered in TN. UPL. (*rupestris* = of rock, living in rocky places). Plates 617 and 618.

**V. vulpina** L. Frost g. Mesic to wet woodlands and thickets of slopes and bottoms. Across KY and TN. Frequent. FAC, FAC+. (*vulpine* = foxlike, of the fox). Plates 619, 620, and 621.

☠ **Wisteria** Nutt. Wisteria. Family Fabaceae. (for Caspar Wistar, 1761–1818, American anatomist of Pennsylvania University). Woody vines climbing by twining; twigs pubescent, often with knobby projections on each side of leaf scar; terminal buds absent; lateral buds 2-scaled or with silvery pubescence; leaf scars alternate, raised, elliptic, with single bundle scar or indistinct; stipules very small and scars inconspicuous; fruit a legume. The bark, pods, and seeds are highly toxic.

1. Legume glabrous; axillary buds silvery-pubescent and blunt; native vines, slender and rarely climbing into the canopy_____ **W. frutescens.**
1. Legume velvety-pubescent; axillary buds smooth to slightly hairy and acute; non-native vines, robust and climbing into the canopy.
   2. Vines twining counterclockwise_____**W. sinensis.**
   2. Vines twining clockwise _____**W. floribunda.**

** **W. floribunda** (Willd.) DC. Japanese w. Thickets and roadsides. Across TN. Lesser Threat in TN. (*sinense* = from China). Plates 622 and 623.

**W. frutescens** (L.) Poir. American w. Alluvial forests, stream banks, sloughs. Across KY and TN. Infrequent. FACW–, FACW. (*frutescens* = shrubby, becoming shrubby). Plates 624 and 625.

* **W. sinensis** (Sims) Sweet. Chinese w. Thickets and roadsides. Across KY and TN. Infrequent, naturalized from cultivation, a native of China. Not pictured.

**Xanthorhiza** Marshall. Family Ranunculaceae. (*Xanthorhiza* = yellow root). Shrub < 1 m tall, the stems usually unbranched, spreading by rhizomes, with roots and inner bark lemon-yellow; terminal buds present, much larger than lateral ones, with about 5 imbricate scales, to

18 mm long; lateral buds appressed, about 3-scaled; leaf scars alternate, crowded toward the stem apex, curved and nearly encircling the twig, with about 10 bundle scars; stipular scars absent; fruit a dry 1-seeded follicle, in drooping clusters.

**X. simplicissima** Marshall. Yellowroot. Mesic forests, stream banks. E KY and E TN, with disjunct populations in W IP of TN. Infrequent. FACW, FACW–. (*simplicissima* = the least divided). Plate 626.

**Yucca** L. Spanish bayonet, Adam's needle. Family Agavaceae. (*Yucca* = based on yuca, an alternate name for cassava or manioc, referring to the similarly enlarged roots). Shrublike with short (to 40 cm) thick stem topped by a cluster of evergreen straplike leaves, these to 80 cm long by 7 cm wide, with parallel venation and fibrous threads along the margin, the leaf terminated by a stiff spine; fruiting stalk emerging from center of leaf cluster, to 3 m tall, bearing cylindric, 6-sectioned capsules, to 4 cm long, with black seeds. The tall fruiting stalks were dried and used as arrow shafts and fire starters by Native Americans. Both species are native to the SE U.S. and have escaped from cultivation in our range, at least in KY.

1. Fruiting branches pubescent _____**Y. flaccida.**
1. Fruiting branches glabrous_____**Y. filamentosa.**

\* **Y. filamentosa** L. Adam's Needle. Barrens, glades, and roadsides. Across KY and TN. (*filamentosa* = thread like, with filaments or thread). Plates 627 and 628.

\* **Y. flaccida** Haw. Yucca. Barrens, glades, and roadsides. Across KY and TN. (*flaccida* = limp, weak, feeble). Plate 629.

**Zanthoxylum** L. Family Rutaceae. (*Zanthoxylum* = yellow wood). Shrub to 8 m tall, all parts with citrus odor; twigs with paired spines at nodes; terminal bud present, red-woolly, the lateral buds smaller; leaf scars alternate, broadly triangular to rounded with 3 bundle scars; fruit a reddish capsule to 5 mm long. This species of *Zanthoxylum* has been called the "toothache tree" because Native Americans applied the chewed bark to relieve toothaches. In addition, a poultice made from the bark was used for treating wounds.

! **Z. americanum** Mill. Common prickly-ash. Dry to mesic, rocky limestone hills. BG of KY and CB of TN, scattered westward in KY and eastward in TN. Infrequent. Special Concern in TN. Plate 630.

# Section V. Plates 1 to 630

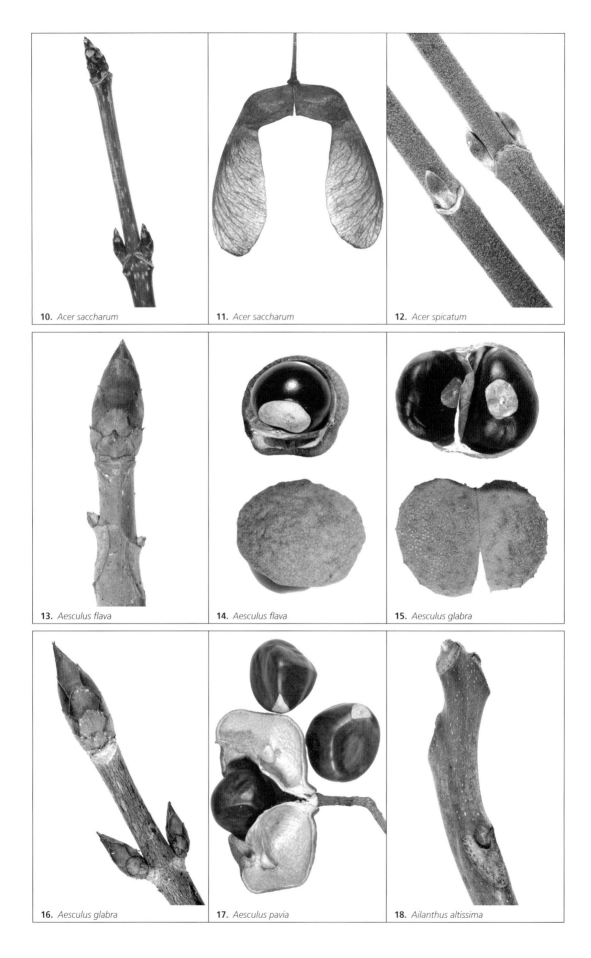

10. *Acer saccharum*
11. *Acer saccharum*
12. *Acer spicatum*
13. *Aesculus flava*
14. *Aesculus flava*
15. *Aesculus glabra*
16. *Aesculus glabra*
17. *Aesculus pavia*
18. *Ailanthus altissima*

19. *Ailanthus altissima*
20. *Akebia quinata*
21. *Albizia julibrissin*
22. *Albizia julibrissin*
23. *Alnus serrulata*
24. *Alnus serrulata*
25. *Alnus serrulata*
26. *Alnus viridis*
27. *Amelanchier arborea*

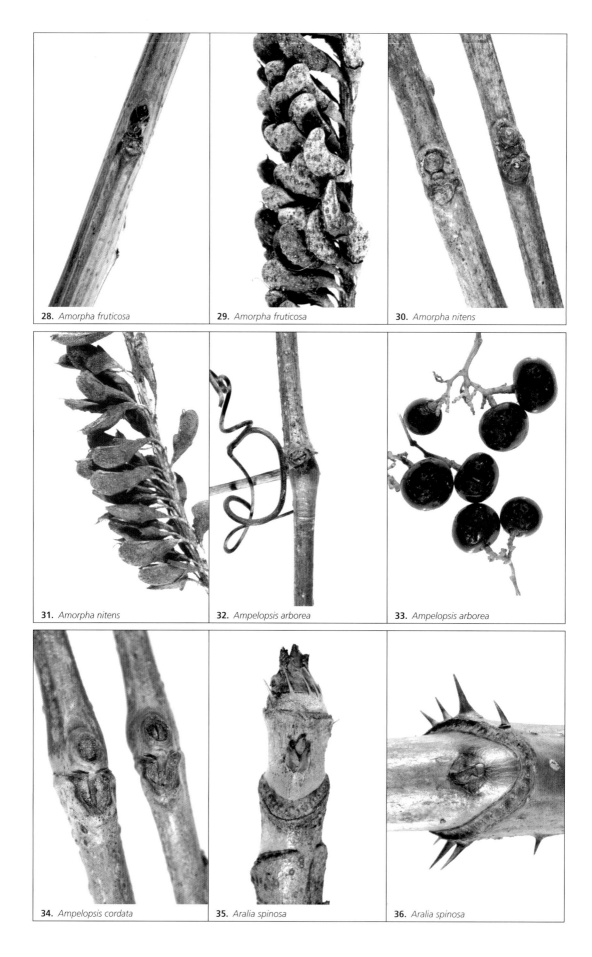

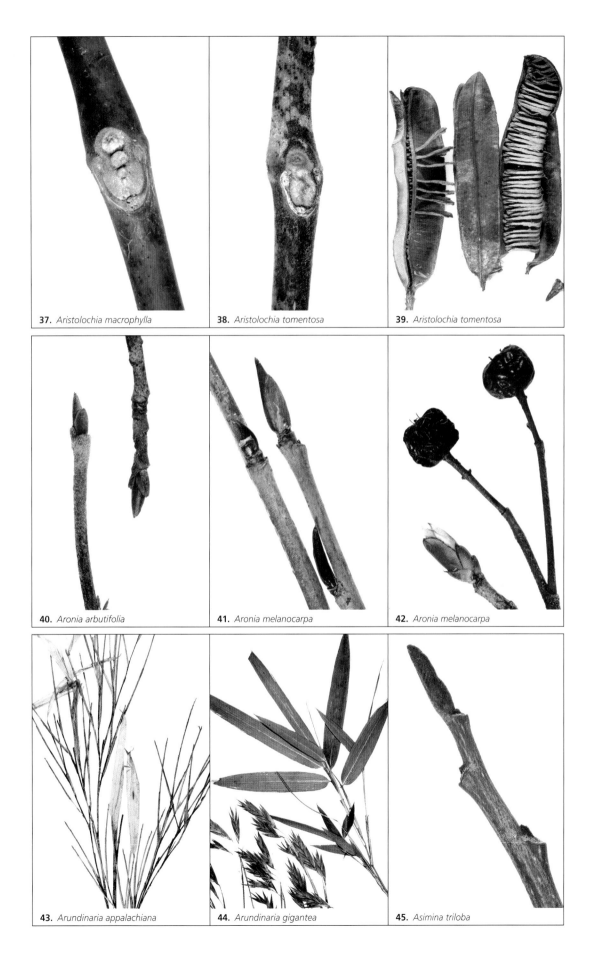

**37.** *Aristolochia macrophylla*  **38.** *Aristolochia tomentosa*  **39.** *Aristolochia tomentosa*

**40.** *Aronia arbutifolia*  **41.** *Aronia melanocarpa*  **42.** *Aronia melanocarpa*

**43.** *Arundinaria appalachiana*  **44.** *Arundinaria gigantea*  **45.** *Asimina triloba*

**46.** *Baccharis halimifolia*  **47.** *Baccharis halimifolia*  **48.** *Berberis canadensis*

**49.** *Berberis thunbergii*  **50.** *Berchemia scandens*  **51.** *Berchemia scandens*

**52.** *Betula alleghaniensis*  **53.** *Betula alleghaniensis*  **54.** *Betula alleghaniensis*

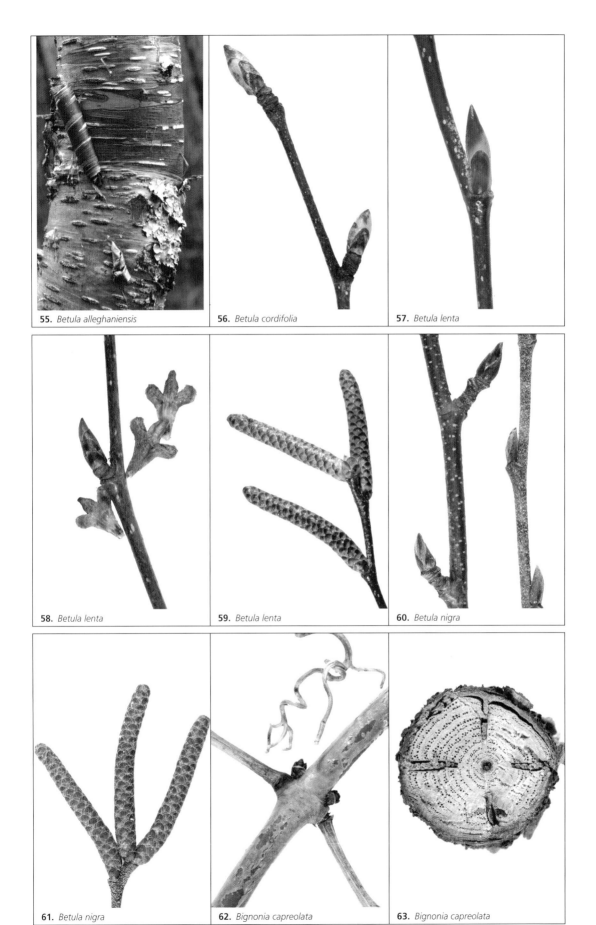

55. *Betula alleghaniensis*
56. *Betula cordifolia*
57. *Betula lenta*
58. *Betula lenta*
59. *Betula lenta*
60. *Betula nigra*
61. *Betula nigra*
62. *Bignonia capreolata*
63. *Bignonia capreolata*

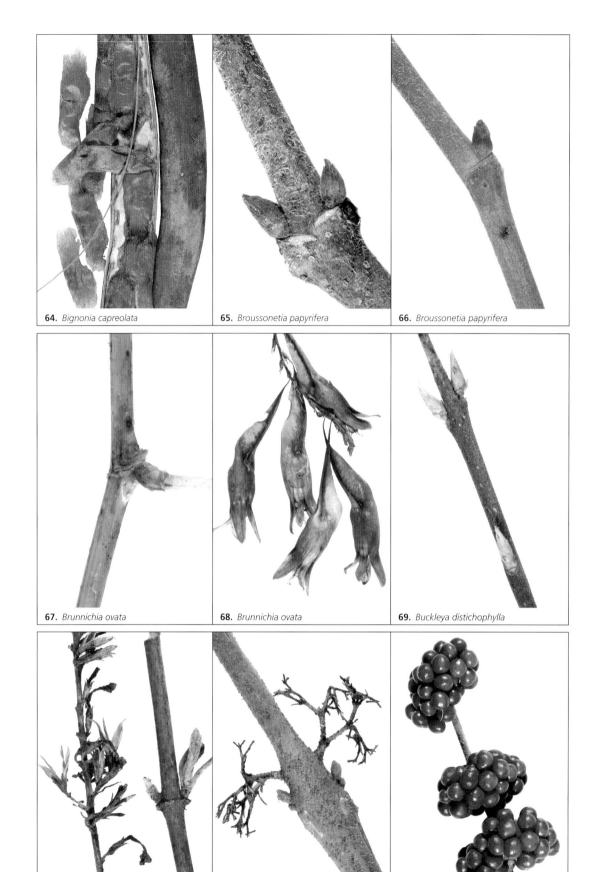

64. *Bignonia capreolata*  65. *Broussonetia papyrifera*  66. *Broussonetia papyrifera*

67. *Brunnichia ovata*  68. *Brunnichia ovata*  69. *Buckleya distichophylla*

70. *Buddleja davidii*  71. *Callicarpa americana*  72. *Callicarpa americana*

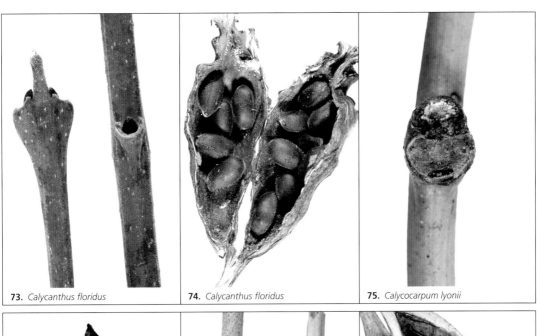

73. *Calycanthus floridus*  74. *Calycanthus floridus*  75. *Calycocarpum lyonii*

76. *Calycocarpum lyonii*  77. *Campsis radicans*  78. *Campsis radicans*

79. *Carpinus caroliniana*  80. *Carya aquatica*  81. *Carya aquatica*

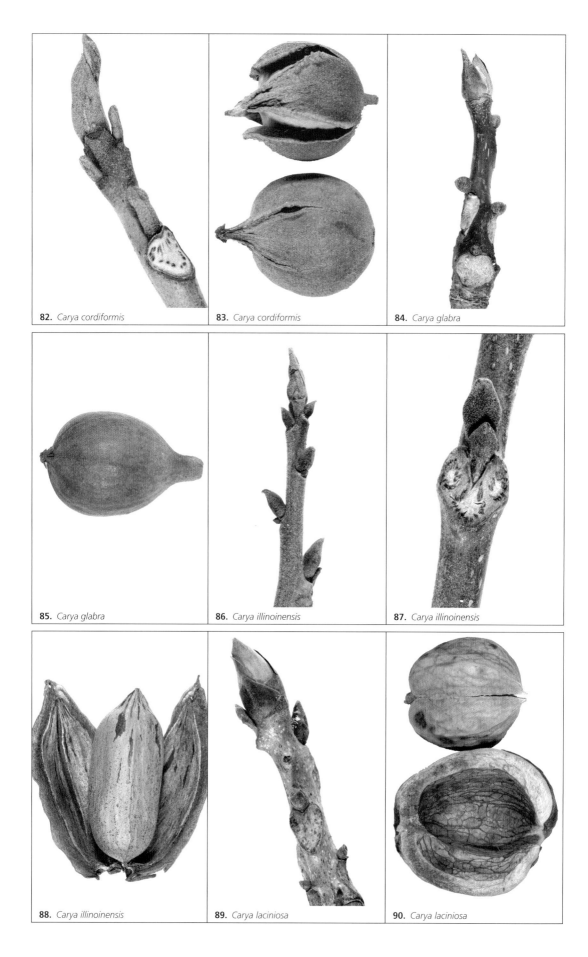

**82.** Carya cordiformis **83.** Carya cordiformis **84.** Carya glabra

**85.** Carya glabra **86.** Carya illinoinensis **87.** Carya illinoinensis

**88.** Carya illinoinensis **89.** Carya laciniosa **90.** Carya laciniosa

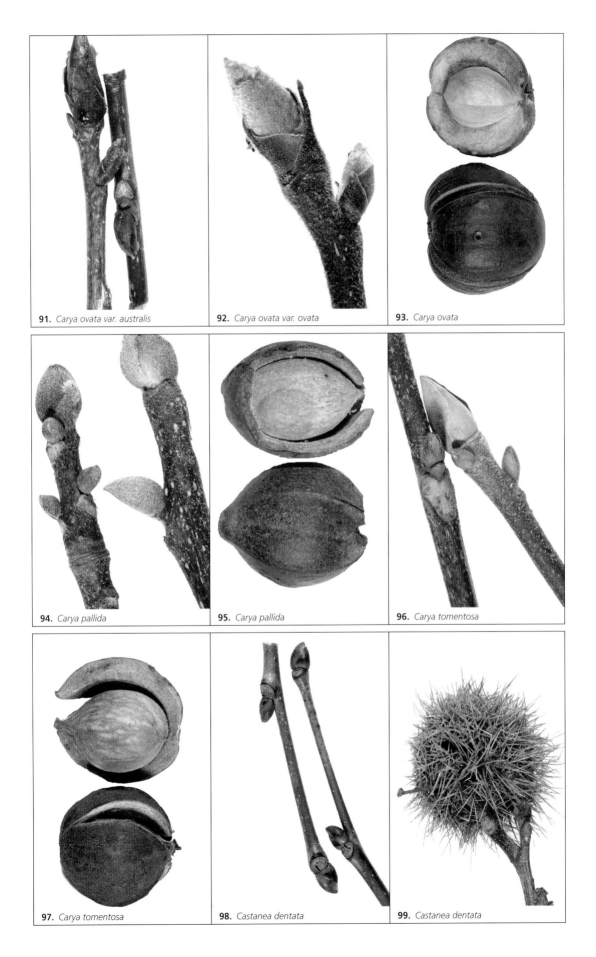

91. *Carya ovata* var. *australis*
92. *Carya ovata* var. *ovata*
93. *Carya ovata*
94. *Carya pallida*
95. *Carya pallida*
96. *Carya tomentosa*
97. *Carya tomentosa*
98. *Castanea dentata*
99. *Castanea dentata*

**100.** *Castanea dentata*  
**101.** *Castanea mollissima*  
**102.** *Castanea mollissima*  
**103.** *Castanea pumila*  
**104.** *Castanea pumila*  
**105.** *Catalpa bignonioides*  
**106.** *Catalpa bignonioides*  
**107.** *Catalpa speciosa*  
**108.** *Catalpa speciosa*

**109.** *Ceanothus americanus*  
**110.** *Ceanothus americanus*  
**111.** *Celastrus orbiculatus*  
**112.** *Celastrus orbiculatus*  
**113.** *Celastrus scandens*  
**114.** *Celtis occidentalis*

**115.** *Celtis tenuifolia*  
**116.** *Cephalanthus occidentalis*  
**117.** *Cercis canadensis*

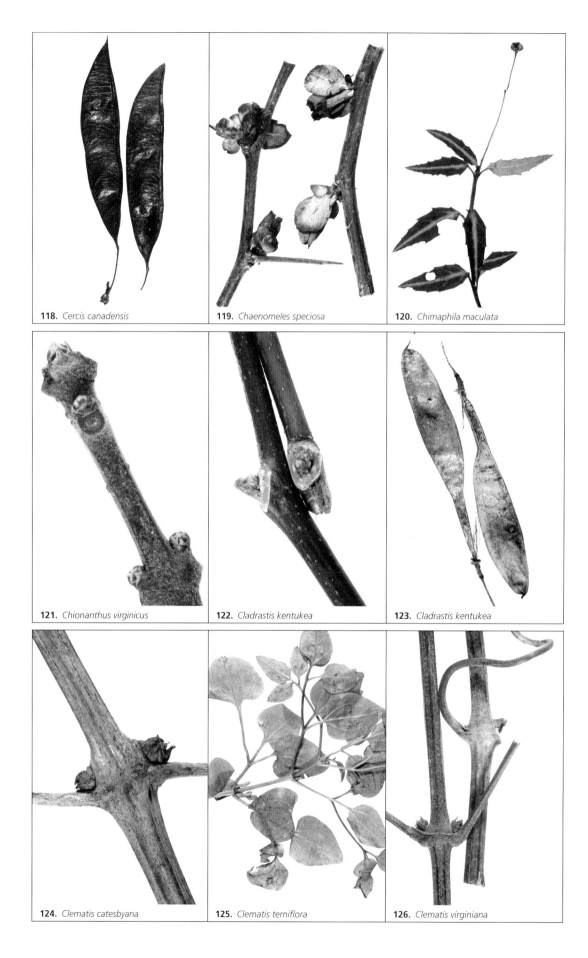

**118.** *Cercis canadensis*  
**119.** *Chaenomeles speciosa*  
**120.** *Chimaphila maculata*  
**121.** *Chionanthus virginicus*  
**122.** *Cladrastis kentukea*  
**123.** *Cladrastis kentukea*  
**124.** *Clematis catesbyana*  
**125.** *Clematis terniflora*  
**126.** *Clematis virginiana*

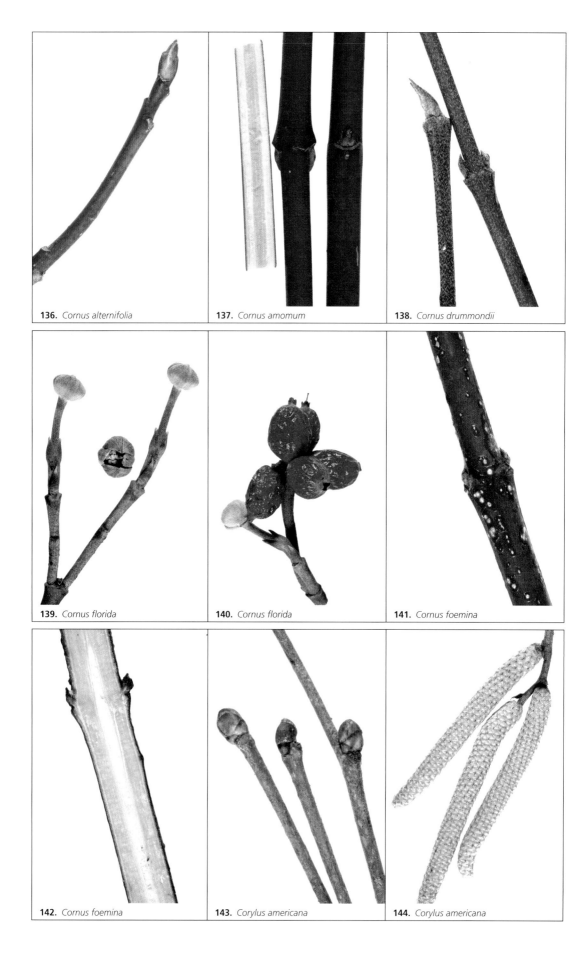

136. *Cornus alternifolia*
137. *Cornus amomum*
138. *Cornus drummondii*
139. *Cornus florida*
140. *Cornus florida*
141. *Cornus foemina*
142. *Cornus foemina*
143. *Corylus americana*
144. *Corylus americana*

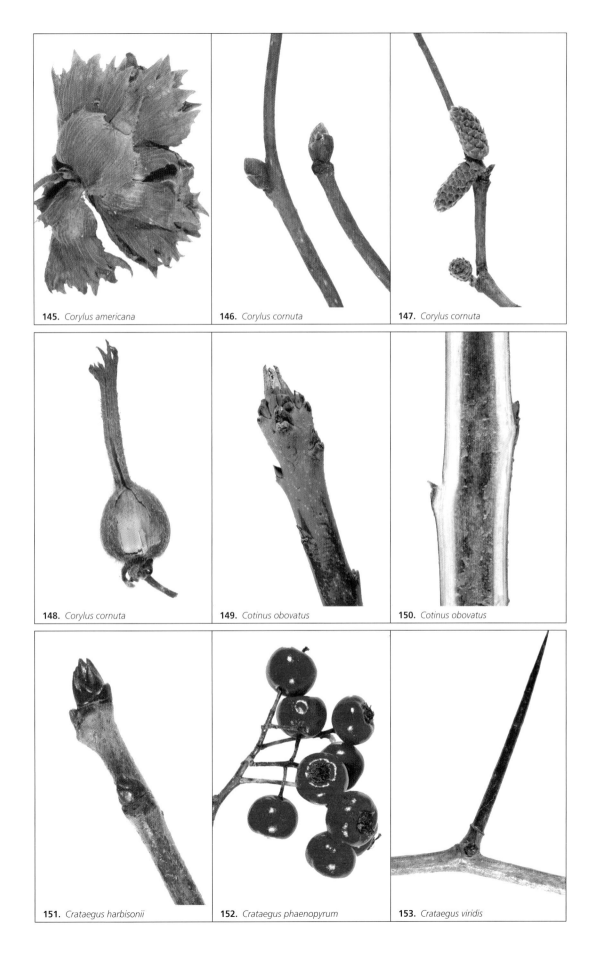

**145.** *Corylus americana*
**146.** *Corylus cornuta*
**147.** *Corylus cornuta*
**148.** *Corylus cornuta*
**149.** *Cotinus obovatus*
**150.** *Cotinus obovatus*
**151.** *Crataegus harbisonii*
**152.** *Crataegus phaenopyrum*
**153.** *Crataegus viridis*

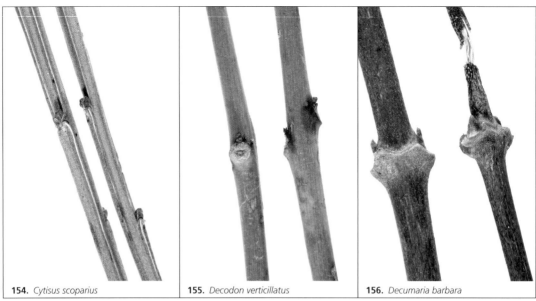

**154.** *Cytisus scoparius*   **155.** *Decodon verticillatus*   **156.** *Decumaria barbara*

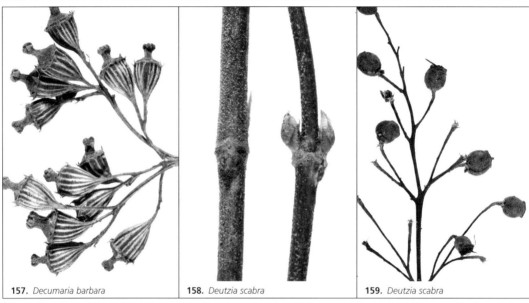

**157.** *Decumaria barbara*   **158.** *Deutzia scabra*   **159.** *Deutzia scabra*

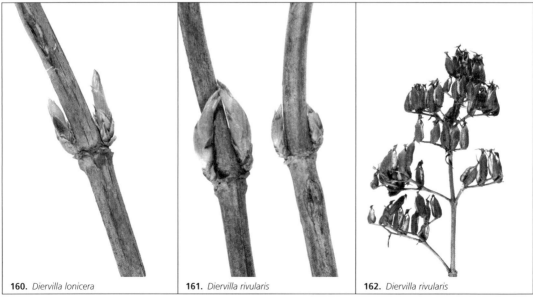

**160.** *Diervilla lonicera*   **161.** *Diervilla rivularis*   **162.** *Diervilla rivularis*

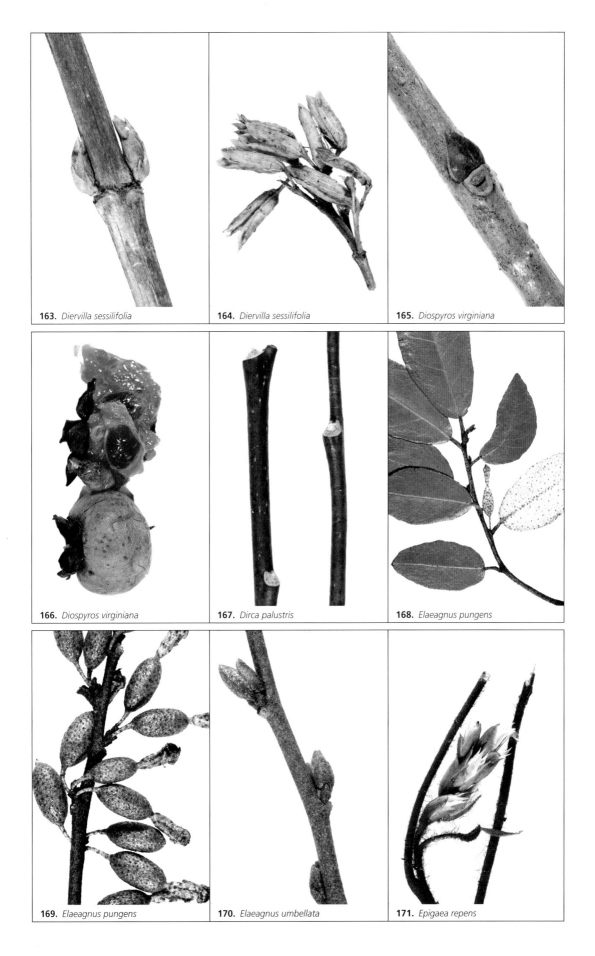

163. *Diervilla sessilifolia*  164. *Diervilla sessilifolia*  165. *Diospyros virginiana*

166. *Diospyros virginiana*  167. *Dirca palustris*  168. *Elaeagnus pungens*

169. *Elaeagnus pungens*  170. *Elaeagnus umbellata*  171. *Epigaea repens*

172. *Epigaea repens*
173. *Euonymus alatus*
174. *Euonymus alatus*
175. *Euonymus americanus*
176. *Euonymus americanus*
177. *Euonymus atropurpureus*
178. *Euonymus atropurpureus*
179. *Euonymus fortunei*
180. *Euonymus obovatus*

181. *Euonymus obovatus*
182. *Fagus grandifolia*
183. *Fagus grandifolia*
184. *Fagus grandifolia*
185. *Forestiera acuminata*
186. *Forestiera ligustrina*
187. *Forsythia viridissima*
188. *Fothergilla major*
189. *Fothergilla major*

**190.** *Frangula caroliniana*  **191.** *Fraxinus americana*  **192.** *Fraxinus americana*

**193.** *Fraxinus pennsylvanica*  **194.** *Fraxinus pennsylvanica*  **195.** *Fraxinus profunda*

**196.** *Fraxinus profunda*  **197.** *Fraxinus quadrangulata*  **198.** *Fraxinus quadrangulata*

**199.** *Gaultheria procumbens*
**200.** *Gaylussacia baccata*
**201.** *Gaylussacia brachycera*
**202.** *Gaylussacia dumosa*
**203.** *Gaylussacia ursina*
**204.** *Gelsemium sempervirens*
**205.** *Gelsemium sempervirens*
**206.** *Gleditsia aquatica*
**207.** *Gleditsia aquatica*

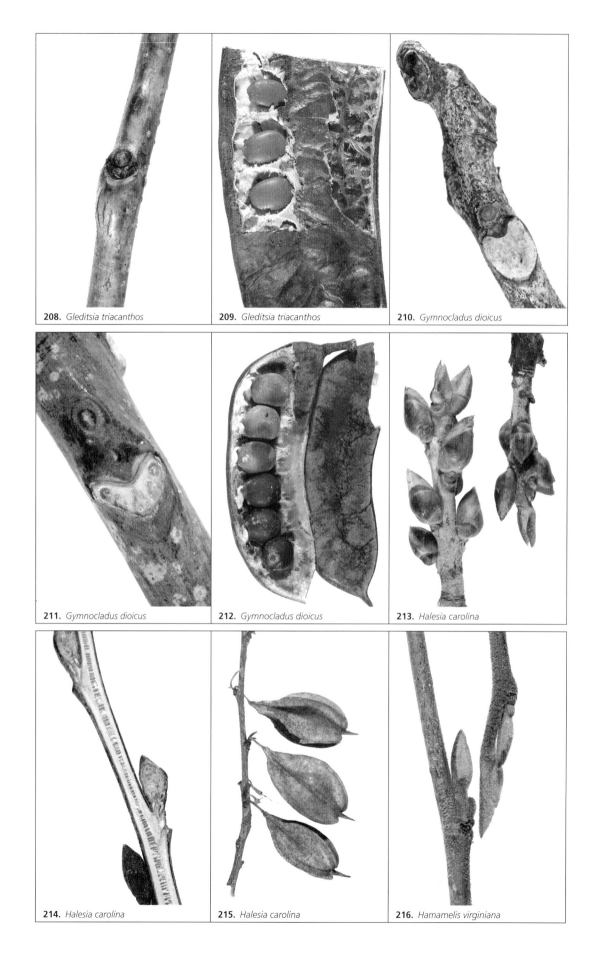

208. *Gleditsia triacanthos*
209. *Gleditsia triacanthos*
210. *Gymnocladus dioicus*
211. *Gymnocladus dioicus*
212. *Gymnocladus dioicus*
213. *Halesia carolina*
214. *Halesia carolina*
215. *Halesia carolina*
216. *Hamamelis virginiana*

**217.** *Hamamelis virginiana*  **218.** *Hedera helix*  **219.** *Hedera helix*

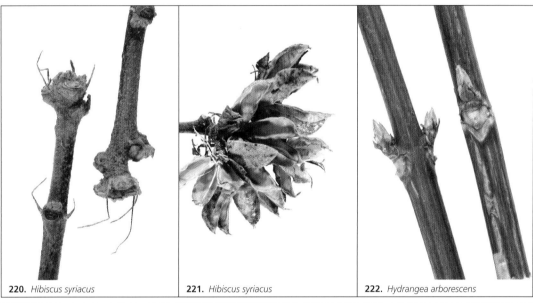

**220.** *Hibiscus syriacus*  **221.** *Hibiscus syriacus*  **222.** *Hydrangea arborescens*

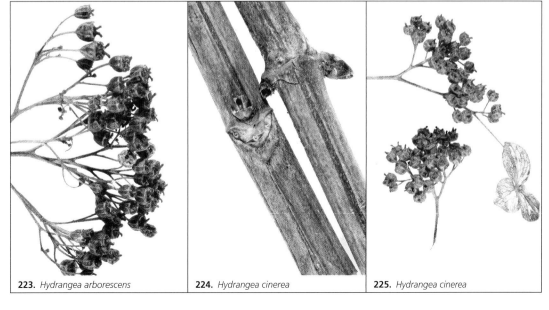

**223.** *Hydrangea arborescens*  **224.** *Hydrangea cinerea*  **225.** *Hydrangea cinerea*

**226.** *Hydrangea quercifolia*
**227.** *Hydrangea quercifolia*
**228.** *Hydrangea radiata*
**229.** *Hypericum crux-andreae*
**230.** *Hypericum crux-andreae*
**231.** *Hypericum densiflorum*
**232.** *Hypericum frondosum*
**233.** *Hypericum hypericoides*
**234.** *Hypericum lobocarpum*

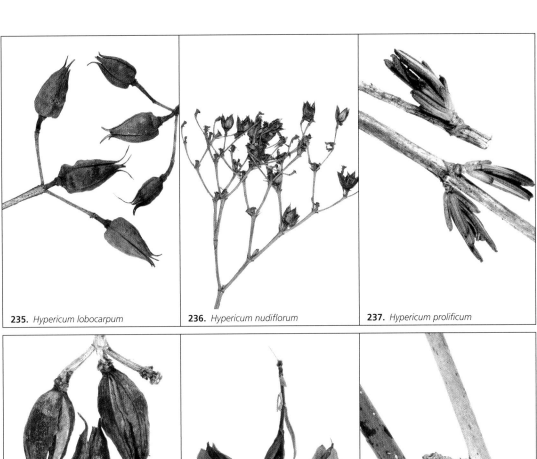

**235.** *Hypericum lobocarpum*  **236.** *Hypericum nudiflorum*  **237.** *Hypericum prolificum*

**238.** *Hypericum prolificum*  **239.** *Hypericum stragulum*  **240.** *Ilex ambigua*

**241.** *Ilex decidua*  **242.** *Ilex decidua*  **243.** *Ilex longipes*

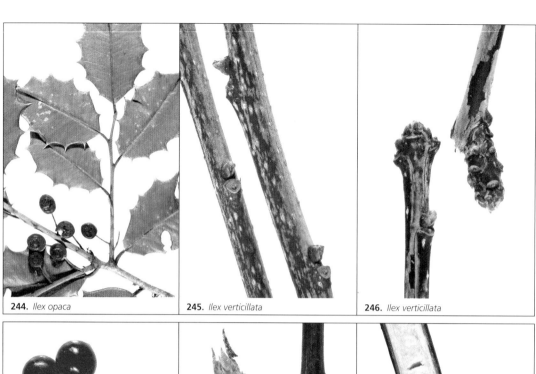

244. *Ilex opaca*  245. *Ilex verticillata*  246. *Ilex verticillata*

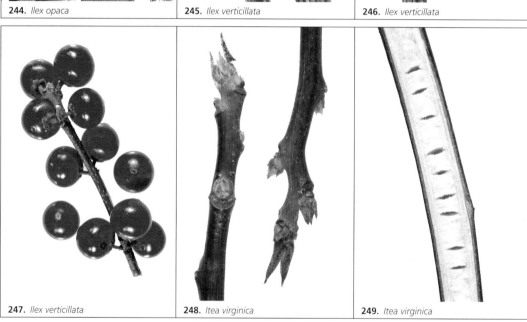

247. *Ilex verticillata*  248. *Itea virginica*  249. *Itea virginica*

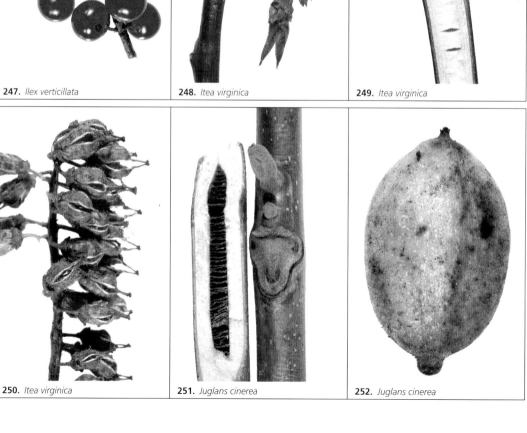

250. *Itea virginica*  251. *Juglans cinerea*  252. *Juglans cinerea*

**253.** *Juglans nigra*
**254.** *Juglans nigra*
**255.** *Juniperus virginiana*
**256.** *Juniperus virginiana*
**257.** *Kalmia carolina*
**258.** *Kalmia latifolia*
**259.** *Kalmia latifolia*
**260.** *Kerria japonica*
**261.** *Kerria japonica*

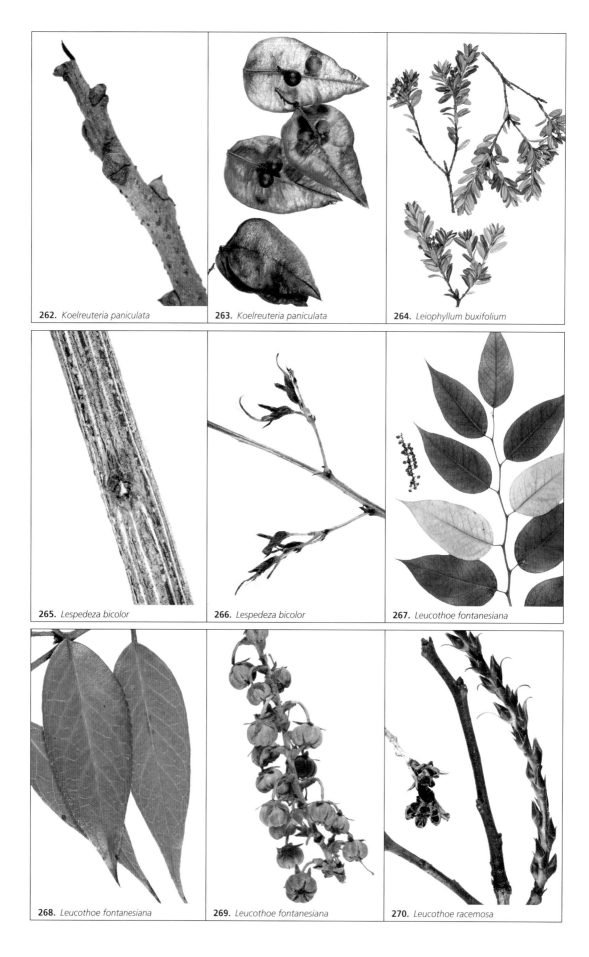

262. *Koelreuteria paniculata*
263. *Koelreuteria paniculata*
264. *Leiophyllum buxifolium*
265. *Lespedeza bicolor*
266. *Lespedeza bicolor*
267. *Leucothoe fontanesiana*
268. *Leucothoe fontanesiana*
269. *Leucothoe fontanesiana*
270. *Leucothoe racemosa*

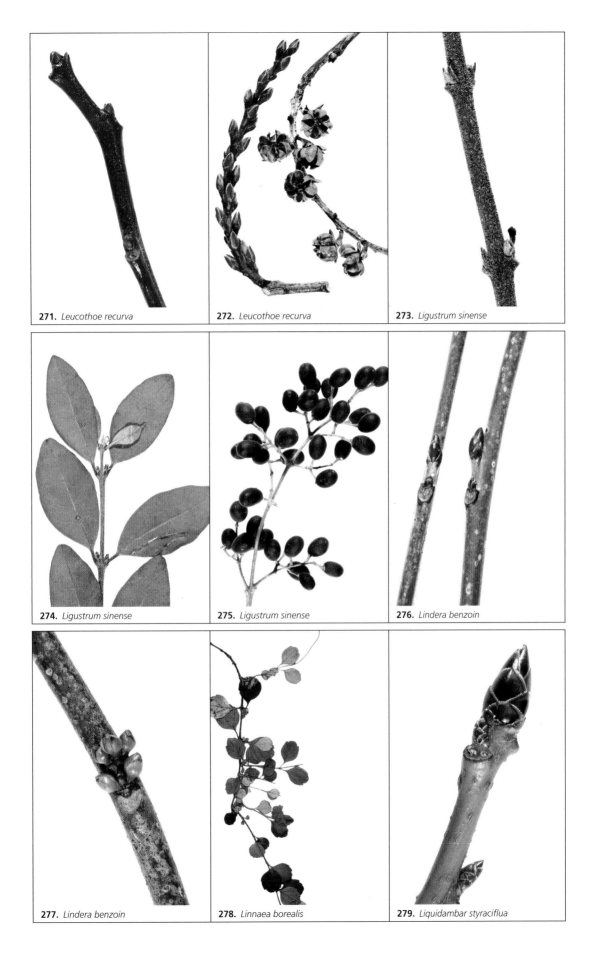

**271.** *Leucothoe recurva*
**272.** *Leucothoe recurva*
**273.** *Ligustrum sinense*
**274.** *Ligustrum sinense*
**275.** *Ligustrum sinense*
**276.** *Lindera benzoin*
**277.** *Lindera benzoin*
**278.** *Linnaea borealis*
**279.** *Liquidambar styraciflua*

280. *Liquidambar styraciflua*
281. *Liriodendron tulipifera*
282. *Liriodendron tulipifera*
283. *Liriodendron tulipifera*
284. *Lonicera dioica*
285. *Lonicera flava*
286. *Lonicera fragrantissima*
287. *Lonicera fragrantissima*
288. *Lonicera japonica*

289. *Lonicera japonica*
290. *Lonicera japonica*
291. *Lonicera japonica*
292. *Lonicera maackii*
293. *Lonicera maackii*
294. *Lonicera maackii*
295. *Lonicera sempervirens*
296. *Lycium barbarum*
297. *Lyonia ligustrina*

298. *Lyonia ligustrina*
299. *Maclura pomifera*
300. *Maclura pomifera*
301. *Maclura pomifera*
302. *Magnolia acuminata*
303. *Magnolia fraseri*
304. *Magnolia fraseri*
305. *Magnolia grandiflora*
306. *Magnolia macrophylla*

307. *Magnolia macrophylla*
308. *Magnolia macrophylla*
309. *Magnolia tripetala*
310. *Magnolia tripetala*
311. *Magnolia virginiana*
312. *Mahonia bealei*
313. *Malus angustifolia*
314. *Malus coronaria*
315. *Malus pumila*

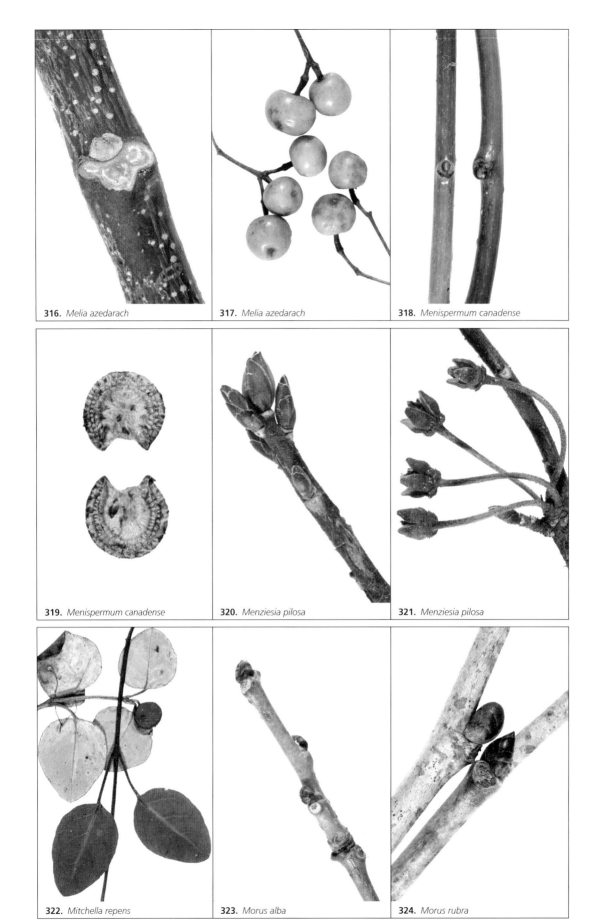

316. *Melia azedarach*
317. *Melia azedarach*
318. *Menispermum canadense*
319. *Menispermum canadense*
320. *Menziesia pilosa*
321. *Menziesia pilosa*
322. *Mitchella repens*
323. *Morus alba*
324. *Morus rubra*

325. *Nandina domestica*
326. *Nandina domestica*
327. *Nemopanthus collinus*
328. *Nestronia umbellula*
329. *Nestronia umbellula*
330. *Neviusia alabamensis*
331. *Neviusia alabamensis*
332. *Nyssa aquatica*
333. *Nyssa biflora*

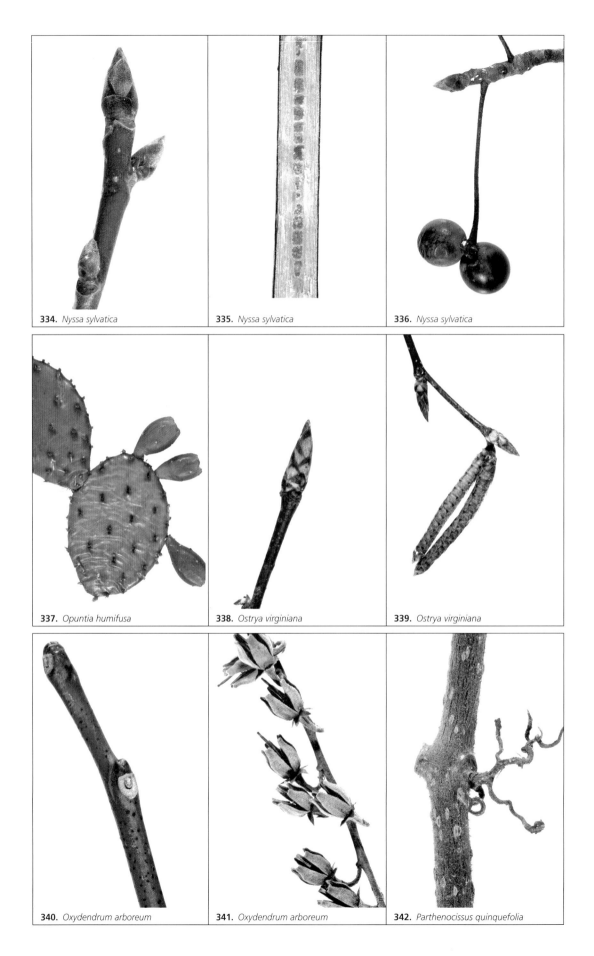

334. *Nyssa sylvatica*
335. *Nyssa sylvatica*
336. *Nyssa sylvatica*
337. *Opuntia humifusa*
338. *Ostrya virginiana*
339. *Ostrya virginiana*
340. *Oxydendrum arboreum*
341. *Oxydendrum arboreum*
342. *Parthenocissus quinquefolia*

343. *Paulownia tomentosa*
344. *Paulownia tomentosa*
345. *Paulownia tomentosa*
346. *Paulownia tomentosa*
347. *Paxistima canbyi*
348. *Philadelphus hirsutus*
349. *Philadelphus inodorus*
350. *Philadelphus inodorus*
351. *Philadelphus pubescens*

**352.** *Phoradendron leucarpum*  
**353.** *Physocarpus opulifolius*  
**354.** *Physocarpus opulifolius*

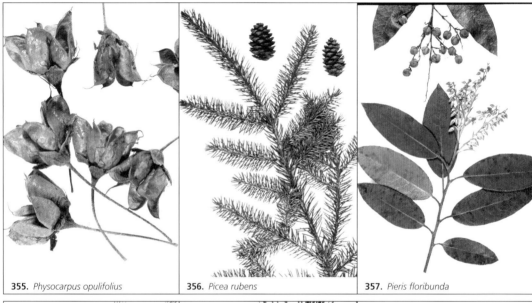

**355.** *Physocarpus opulifolius*  
**356.** *Picea rubens*  
**357.** *Pieris floribunda*

**358.** *Pinus echinata*  
**359.** *Pinus pungens*  
**360.** *Pinus rigida*

**361.** *Pinus strobus*  **362.** *Pinus taeda*  **363.** *Pinus virginiana*

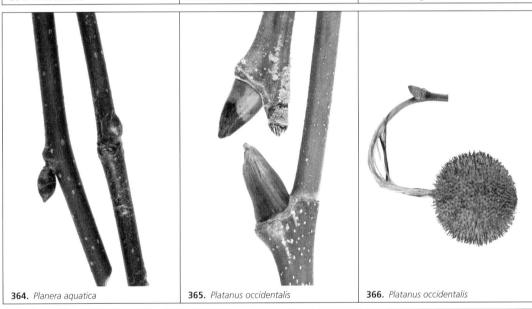

**364.** *Planera aquatica*  **365.** *Platanus occidentalis*  **366.** *Platanus occidentalis*

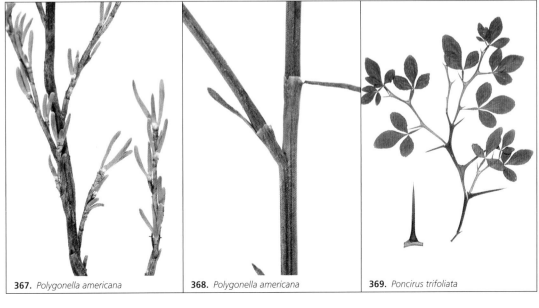

**367.** *Polygonella americana*  **368.** *Polygonella americana*  **369.** *Poncirus trifoliata*

370. *Populus alba*
371. *Populus deltoides*
372. *Populus grandidentata*
373. *Populus heterophylla*
374. *Populus × jackii*
375. *Prunus americana*
376. *Prunus angustifolia*
377. *Prunus cerasus*
378. *Prunus mahaleb*

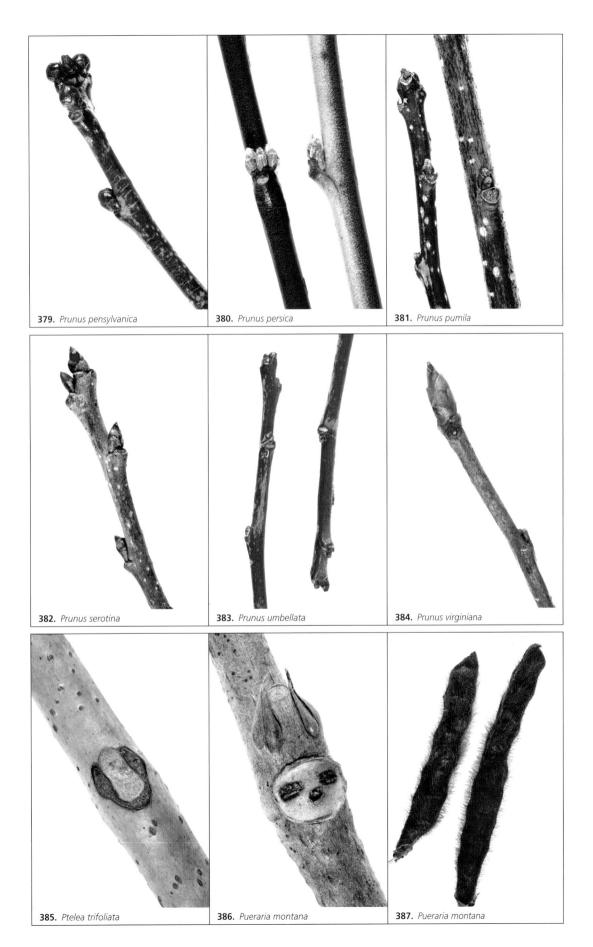

379. *Prunus pensylvanica*
380. *Prunus persica*
381. *Prunus pumila*
382. *Prunus serotina*
383. *Prunus umbellata*
384. *Prunus virginiana*
385. *Ptelea trifoliata*
386. *Pueraria montana*
387. *Pueraria montana*

388. *Pyrularia pubera*
389. *Pyrus calleryana*
390. *Pyrus calleryana*
391. *Pyrus calleryana*
392. *Pyrus communis*
393. *Quercus alba*
394. *Quercus alba*
395. *Quercus bicolor*
396. *Quercus bicolor*

**397.** *Quercus coccinea*  **398.** *Quercus coccinea*  **399.** *Quercus falcata*

**400.** *Quercus falcata*  **401.** *Quercus imbricaria*  **402.** *Quercus imbricaria*

**403.** *Quercus lyrata*  **404.** *Quercus lyrata*  **405.** *Quercus macrocarpa*

**406.** *Quercus macrocarpa*
**407.** *Quercus margaretta*
**408.** *Quercus marilandica*
**409.** *Quercus marilandica*
**410.** *Quercus michauxii*
**411.** *Quercus michauxii*
**412.** *Quercus montana*
**413.** *Quercus montana*
**414.** *Quercus muhlenbergii*

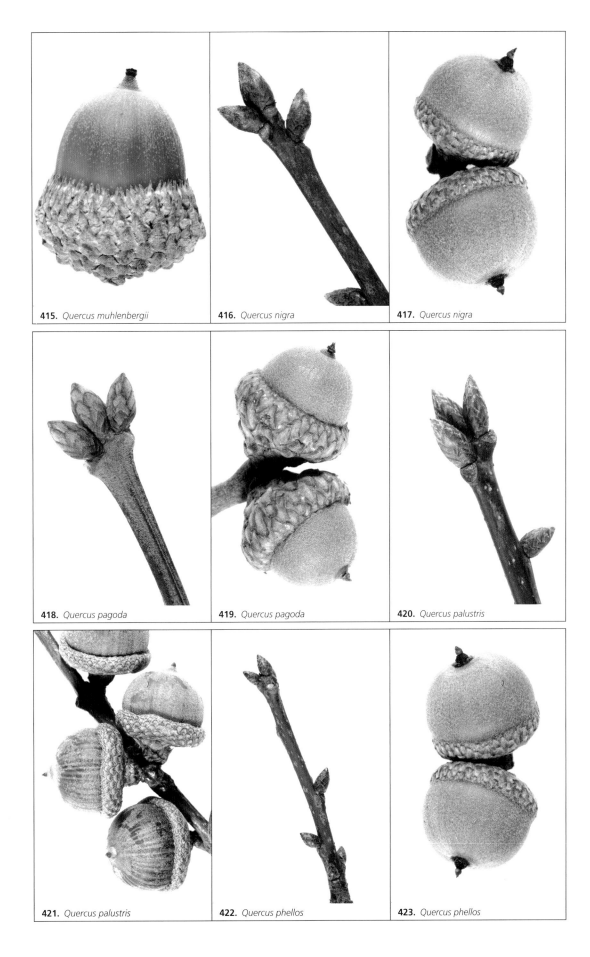

**415.** *Quercus muhlenbergii*  **416.** *Quercus nigra*  **417.** *Quercus nigra*
**418.** *Quercus pagoda*  **419.** *Quercus pagoda*  **420.** *Quercus palustris*
**421.** *Quercus palustris*  **422.** *Quercus phellos*  **423.** *Quercus phellos*

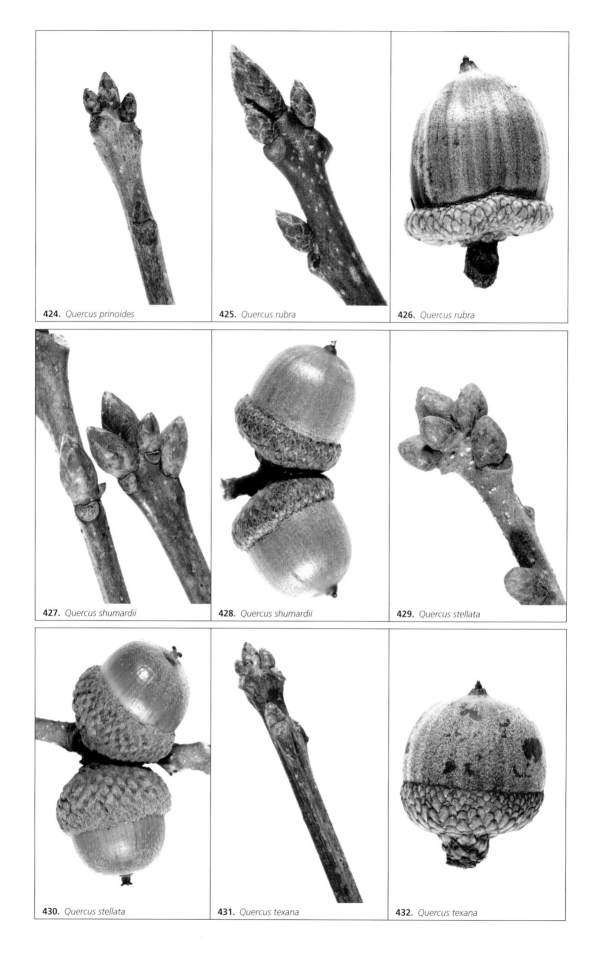

**424.** *Quercus prinoides*
**425.** *Quercus rubra*
**426.** *Quercus rubra*
**427.** *Quercus shumardii*
**428.** *Quercus shumardii*
**429.** *Quercus stellata*
**430.** *Quercus stellata*
**431.** *Quercus texana*
**432.** *Quercus texana*

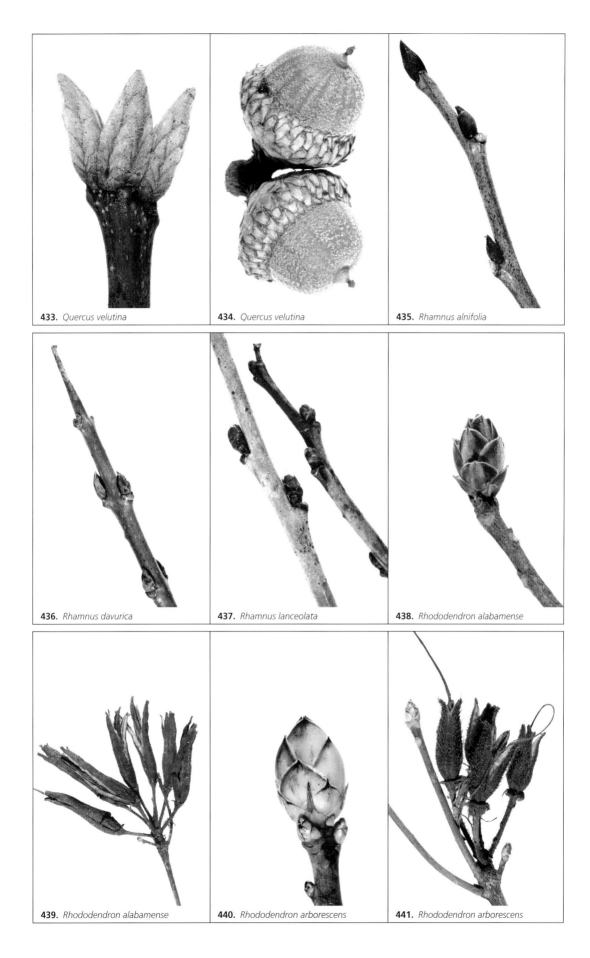

**433.** *Quercus velutina*  
**434.** *Quercus velutina*  
**435.** *Rhamnus alnifolia*  
**436.** *Rhamnus davurica*  
**437.** *Rhamnus lanceolata*  
**438.** *Rhododendron alabamense*  
**439.** *Rhododendron alabamense*  
**440.** *Rhododendron arborescens*  
**441.** *Rhododendron arborescens*

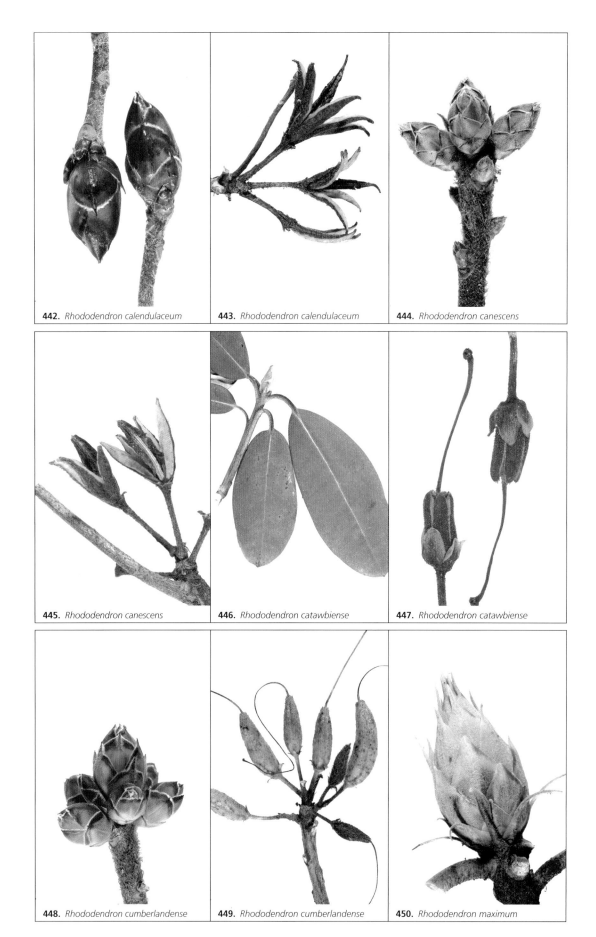

442. *Rhododendron calendulaceum*
443. *Rhododendron calendulaceum*
444. *Rhododendron canescens*
445. *Rhododendron canescens*
446. *Rhododendron catawbiense*
447. *Rhododendron catawbiense*
448. *Rhododendron cumberlandense*
449. *Rhododendron cumberlandense*
450. *Rhododendron maximum*

451. *Rhododendron maximum*
452. *Rhododendron minus*
453. *Rhododendron periclymenoides*
454. *Rhododendron periclymenoides*
455. *Rhododendron viscosum*
456. *Rhodotypos scandens*
457. *Rhodotypos scandens*
458. *Rhus aromatica*
459. *Rhus copallinum*

460. Rhus glabra
461. Rhus glabra
462. Rhus typhina
463. Rhus typhina
464. Ribes aureum
465. Ribes curvatum
466. Ribes curvatum
467. Ribes cynosbati
468. Ribes glandulosum

469. *Ribes missouriense*
470. *Ribes rotundifolium*
471. *Robinia hispida*
472. *Robinia pseudoacacia*
473. *Robinia pseudoacacia*
474. *Robinia pseudoacacia*
475. *Rosa carolina*
476. *Rosa carolina*
477. *Rosa multiflora*

478. *Rosa multiflora*  479. *Rosa palustris*  480. *Rosa palustris*

481. *Rosa setigera*  482. *Rosa setigera*  483. *Rosa virginiana*

484. *Rosa virginiana*  485. *Rubus allegheniensis*  486. *Rubus argutus*

487. *Rubus argutus*
488. *Rubus bifrons*
489. *Rubus bifrons*
490. *Rubus canadensis*
491. *Rubus flagellaris*
492. *Rubus hispidus*
493. *Rubus idaeus*
494. *Rubus idaeus*
495. *Rubus longii*

**496.** *Rubus longii*
**497.** *Rubus occidentalis*
**498.** *Rubus odoratus*
**499.** *Rubus phoenicolasius*
**500.** *Rubus trivialis*
**501.** *Salix alba*
**502.** *Salix babylonica*
**503.** *Salix babylonica*
**504.** *Salix caroliniana*

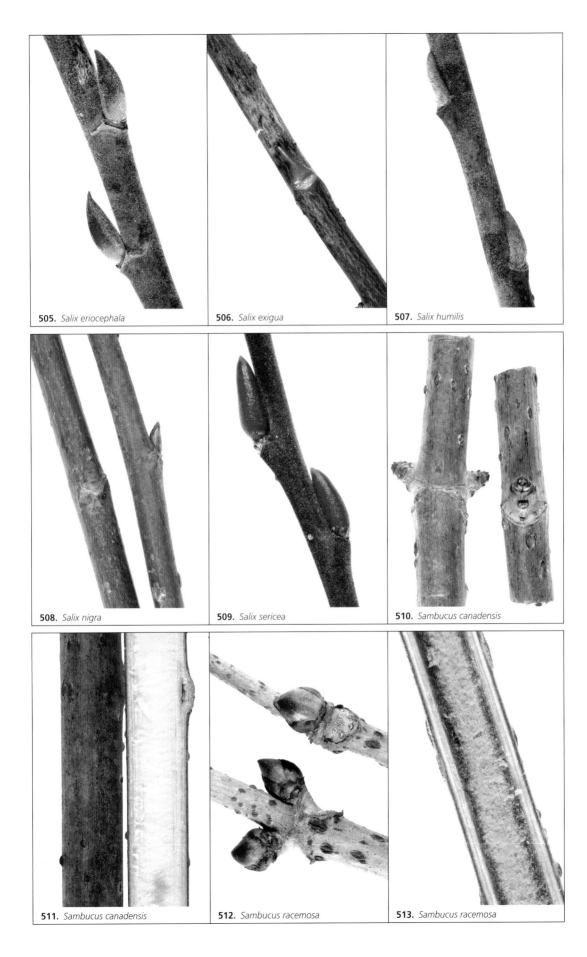

505. *Salix eriocephala*
506. *Salix exigua*
507. *Salix humilis*
508. *Salix nigra*
509. *Salix sericea*
510. *Sambucus canadensis*
511. *Sambucus canadensis*
512. *Sambucus racemosa*
513. *Sambucus racemosa*

514. *Sassafras albidum*  
515. *Schisandra glabra*  
516. *Sibbaldiopsis tridentata*

517. *Sideroxylon lycioides*  
518. *Sideroxylon lycioides*  
519. *Smilax bona-nox*

520. *Smilax bona-nox*  
521. *Smilax glauca*  
522. *Smilax glauca*

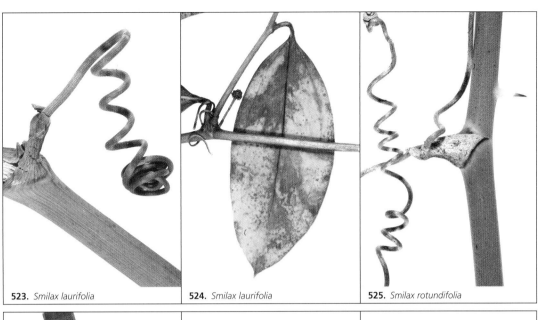

**523.** *Smilax laurifolia*  **524.** *Smilax laurifolia*  **525.** *Smilax rotundifolia*

**526.** *Smilax rotundifolia*  **527.** *Smilax rotundifolia*  **528.** *Smilax tamnoides*

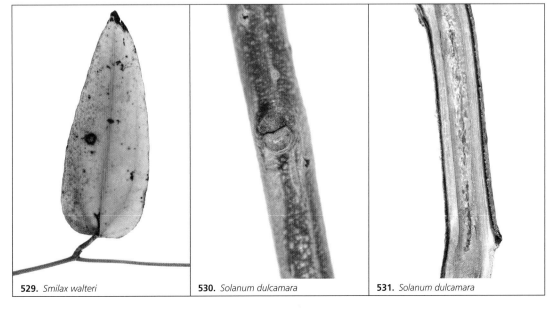

**529.** *Smilax walteri*  **530.** *Solanum dulcamara*  **531.** *Solanum dulcamara*

**532.** *Sorbus americana*  **533.** *Sorbus americana*  **534.** *Spiraea alba*

**535.** *Spiraea alba*  **536.** *Spiraea japonica*  **537.** *Spiraea japonica*

**538.** *Spiraea prunifolia*  **539.** *Spiraea tomentosa*  **540.** *Spiraea tomentosa*

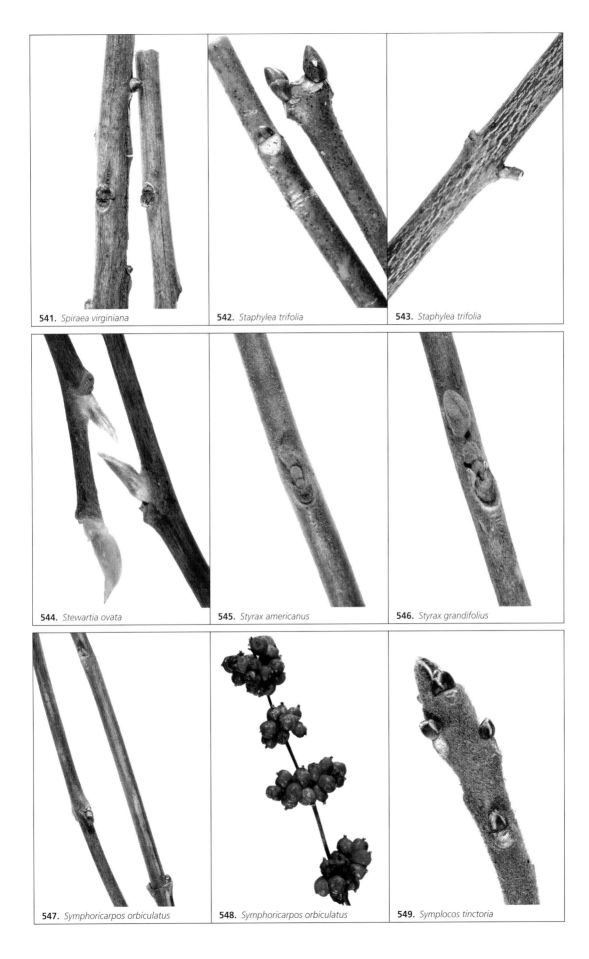

**541.** *Spiraea virginiana*  
**542.** *Staphylea trifolia*  
**543.** *Staphylea trifolia*  
**544.** *Stewartia ovata*  
**545.** *Styrax americanus*  
**546.** *Styrax grandifolius*  
**547.** *Symphoricarpos orbiculatus*  
**548.** *Symphoricarpos orbiculatus*  
**549.** *Symplocos tinctoria*

**550.** *Symplocos tinctoria*   **551.** *Symplocos tinctoria*   **552.** *Syringa vulgaris*

**553.** *Taxodium distichum*   **554.** *Taxodium distichum*   **555.** *Taxus canadensis*

**556.** *Taxus canadensis*   **557.** *Thuja occidentalis*   **558.** *Tilia americana*

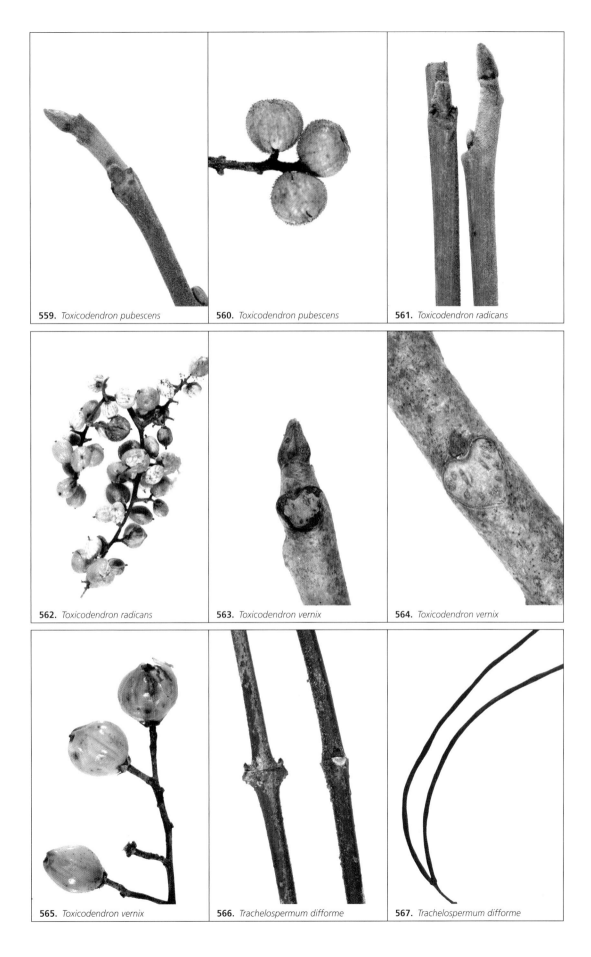

559. *Toxicodendron pubescens*
560. *Toxicodendron pubescens*
561. *Toxicodendron radicans*
562. *Toxicodendron radicans*
563. *Toxicodendron vernix*
564. *Toxicodendron vernix*
565. *Toxicodendron vernix*
566. *Trachelospermum difforme*
567. *Trachelospermum difforme*

568. *Tsuga canadensis*
569. *Tsuga canadensis*
570. *Tsuga caroliniana*
571. *Ulmus alata*
572. *Ulmus alata*
573. *Ulmus americana*
574. *Ulmus crassifolia*
575. *Ulmus rubra*
576. *Ulmus serotina*

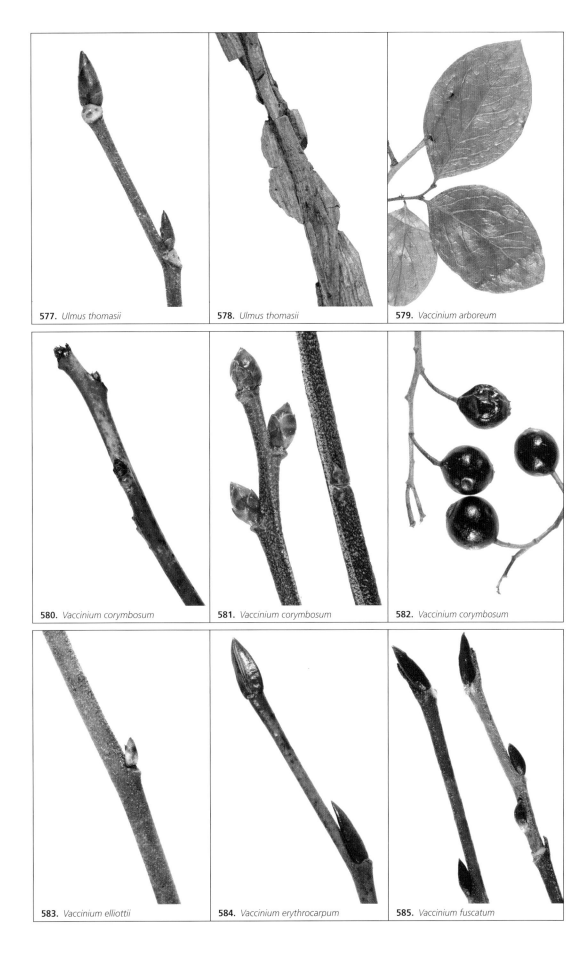

577. *Ulmus thomasii*
578. *Ulmus thomasii*
579. *Vaccinium arboreum*
580. *Vaccinium corymbosum*
581. *Vaccinium corymbosum*
582. *Vaccinium corymbosum*
583. *Vaccinium elliottii*
584. *Vaccinium erythrocarpum*
585. *Vaccinium fuscatum*

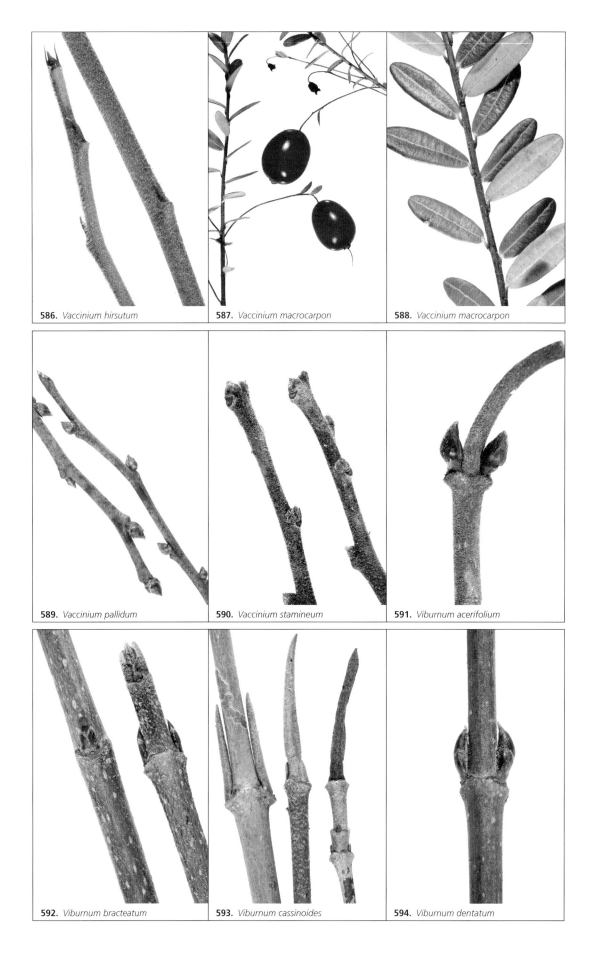

586. *Vaccinium hirsutum*
587. *Vaccinium macrocarpon*
588. *Vaccinium macrocarpon*
589. *Vaccinium pallidum*
590. *Vaccinium stamineum*
591. *Viburnum acerifolium*
592. *Viburnum bracteatum*
593. *Viburnum cassinoides*
594. *Viburnum dentatum*

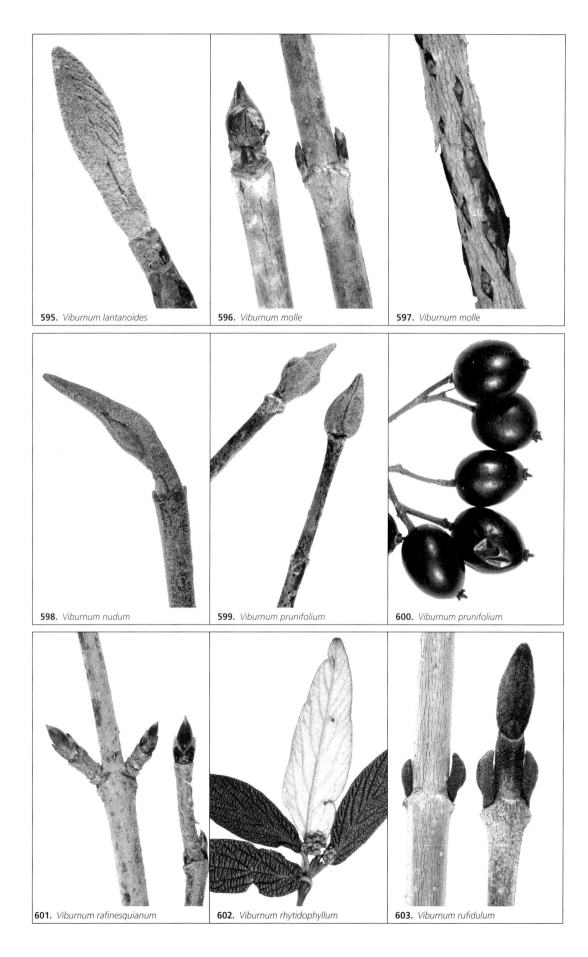

595. *Viburnum lantanoides*
596. *Viburnum molle*
597. *Viburnum molle*
598. *Viburnum nudum*
599. *Viburnum prunifolium*
600. *Viburnum prunifolium*
601. *Viburnum rafinesquianum*
602. *Viburnum rhytidophyllum*
603. *Viburnum rufidulum*

**604.** *Viburnum rufidulum*  **605.** *Vinca major*  **606.** *Vinca minor*

**607.** *Vitis aestivalis*  **608.** *Vitis aestivalis*  **609.** *Vitis cinerea*

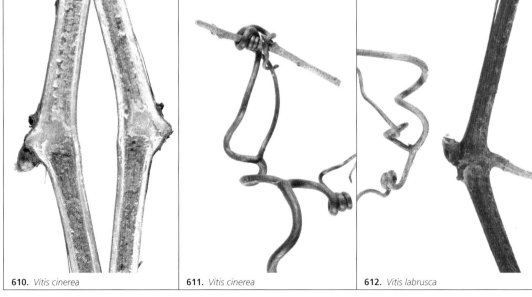

**610.** *Vitis cinerea*  **611.** *Vitis cinerea*  **612.** *Vitis labrusca*

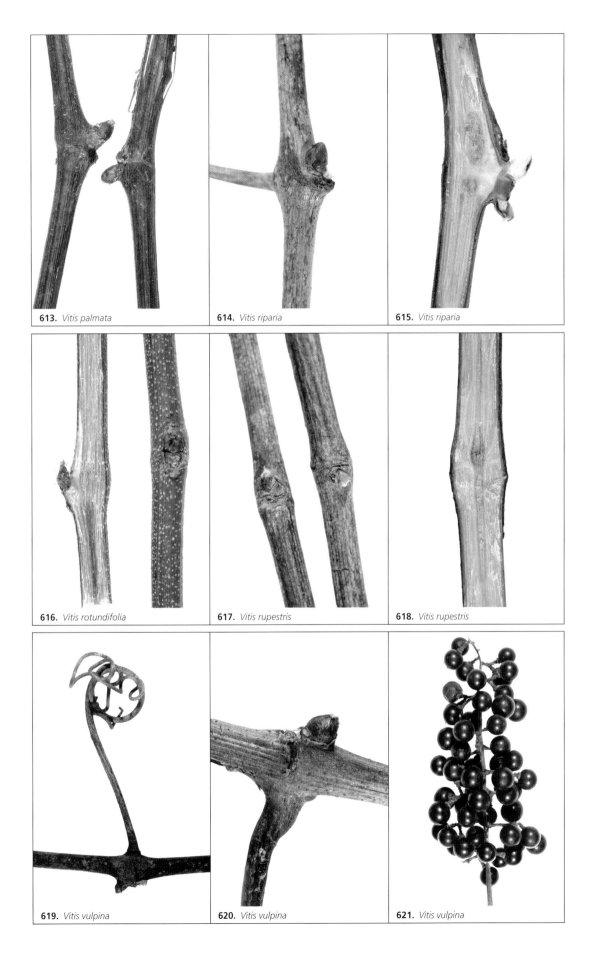

613. *Vitis palmata*
614. *Vitis riparia*
615. *Vitis riparia*
616. *Vitis rotundifolia*
617. *Vitis rupestris*
618. *Vitis rupestris*
619. *Vitis vulpina*
620. *Vitis vulpina*
621. *Vitis vulpina*

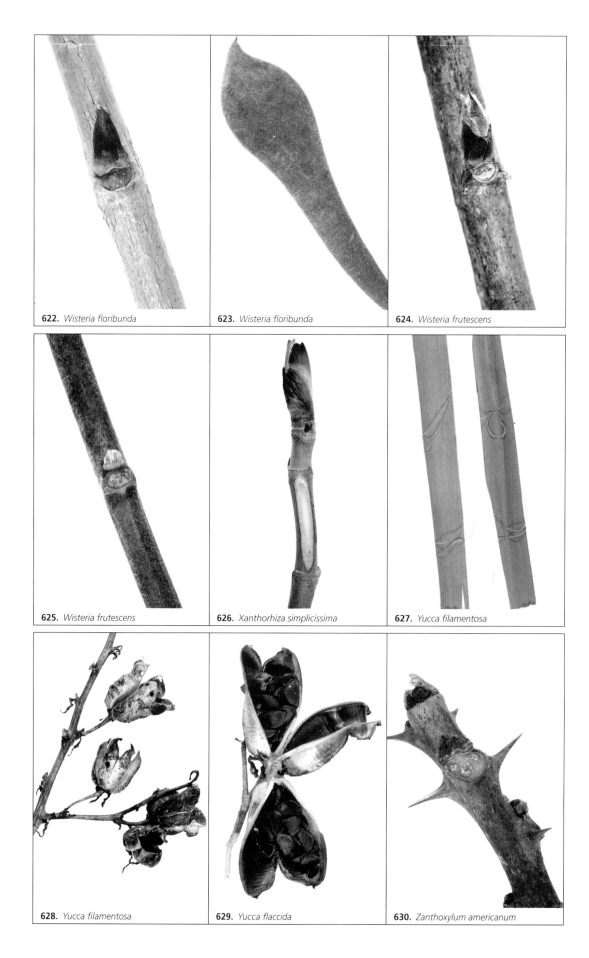

622. *Wisteria floribunda*
623. *Wisteria floribunda*
624. *Wisteria frutescens*
625. *Wisteria frutescens*
626. *Xanthorhiza simplicissima*
627. *Yucca filamentosa*
628. *Yucca filamentosa*
629. *Yucca flaccida*
630. *Zanthoxylum americanum*

# Literature Cited

Brown, R. G., & M. L. Brown. 1972. *Woody plants of Maryland*. University of Maryland Book Center, College Park.

Brummitt, R. K., & C. E. Powell, eds. 1992. *Authors of plant names*. Royal Botanic Gardens, Kew, England.

Bruton, M. S., & D. Estes. 2009. *Baccharis halimifolia* (Asteraceae) is reported as new for the state of Kentucky. *Southeastern Biology* 56: 233.

Burrows, G. E., & R. J. Tyrl. 2001. *Toxic plants of North America*. Iowa State University Press, Ames.

Campbell, C. S., F. Hyland, & M. L. F. Campbell. 1975. *Winter key to woody plants of Maine*. University of Maine Press, Orono.

Chester, E. W., B. E. Wofford, D. Estes, & C. Bailey. 2009. A fifth checklist of Tennessee vascular plants. *Sida, Botanical Miscellany*, No. 31.

Chester, E. W., B. E. Wofford, & R. Kral. 1997. *Atlas of Tennessee vascular plants*. Vol. 2. Misc. Publ. 13. Center for Field Biology, Austin Peay State University, Clarksville, Tennessee.

Chester, E. W., B. E. Wofford, R. Kral, H. R. DeSelm, & A. M. Evans. 1993. *Atlas of Tennessee vascular plants*. Vol. 1. Misc. Publ. 9. Center for Field Biology, Austin Peay State University, Clarksville, Tennessee.

Clark, R. C., & T. J. Weckman. 2008. Annotated catalog and atlas of Kentucky woody plants. *Castanea, Occasional Papers in Eastern Botany*, No. 3.

Cope, E. A. 2001. *Muenscher's keys to woody plants: An expanded guide to native and cultivated species*. Cornell University Press, Ithaca, New York.

Core, E. L., & N. P. Ammons. 1958. *Woody plants in winter*. The Boxwood Press, Pittsburgh, Pennsylvania.

Crabtree, T. 2012. Tennessee Natural Heritage Program. Rare plant list. http://tn.gov/environment/na/pdf/plant_list.pdf.

Department of the Army. 2003. *The illustrated guide to edible wild plants*. The Lyons Press, Guilford, Connecticut.

Elias, T. S., & P. A. Dykeman. 1982. *Field guide to North American edible wild plants*. Outdoor Life Books, Grolier Book Clubs, New York.

Gledhill, D. 2008. *The names of plants*, 4th ed. Cambridge University Press, New York.

Gorman, C. E., M. S. Bruton, and L. D. Estes. 2011. The status of *Elaeagnus multiflora* (Elaeagnaceae), a potentially invasive Asiatic shrub, in Tennessee. *Southeastern Biology* 58: 273.

Harlow, W. H. 1946. *Fruit key and twig key to trees and shrubs*. Dover Publications, Inc., New York.

Jones, R. L. 2005. *Plant life of Kentucky, an illustrated guide to the vascular flora*. University Press of Kentucky, Lexington.

Krochmal, A., & C. Krochmal. 1984. *A field guide to medicinal plants*. Quadrangle/The New York Times Book Company, New York.

[KSNPC] Kentucky State Nature Preserves Commission. 2010. Rare and extirpated biota and natural communities of Kentucky. *J. Ky. Acad. Sci*. 71: 67–81.

[KY-EPPC] Kentucky Exotic Pest Plant Council. 2011. Exotic plant list. http://www.se-eppc.org/ky/list.htm.

Lance, R. 2004. *Woody plants of the southeastern United States, a winter guide*. The University of Georgia Press, Athens.

Lewis, W. M., & M. P. F. Elvin-Lewis. 1977. *Medical botany, plants affecting man's health*. John Wiley and Sons, New York.

Meuninck, J. 2007. *Edible wild plants and useful herbs*. Globe Pequot Press, Guilford, Connecticut.

Meuninck, J. 2008. *Medicinal plants of North America*. Globe Pequot Press, Guilford, Connecticut.

Moerman, D. E. 1998. *Native American ethnobotany*. Timber Press, Portland, Oregon.

Muenscher, W. C. 1975. *Poisonous plants of the United States*. Macmillan Publishing Company, New York.

Peterson, L. A. 1977. *Edible wild plants, eastern/central North America*. Houghton Mifflin Company, New York, New York.

Preston, R. J., Jr., & V. G. Wright. 1978. *Identification of southeastern trees in winter*. North Carolina Agricultural Extension Service, Raleigh.

Swanson, R. E. 1994. *A field guide to the trees and shrubs of the southern mountains*. The John Hopkins University Press, Baltimore, Maryland.

# LITERATURE CITED

Thayer, S. 2006. *The forager's harvest, a guide to identifying, harvesting, and preparing edible wild plants*. Forager's Harvest, Birchwood, Wisconsin.

[TN-EPPC] Tennessee Exotic Pest Plant Council. 2009. Invasive plants of Tennessee. http://www.tneppc.org/invasive_plants.

USDA, NRCS. 2011. The PLANTS Database (http://plants.usda.gov, 9 March 2011). National Plant Data Center, Baton Rouge, LA 70874-4490 USA.

Westbrooks, R. G., & J. W. Preacher. 1986. *Poisonous plants of eastern North America*. University of South Carolina Press, Columbia.

Wofford, B. E., & E. W. Chester. 2002. *Guide to the trees, shrubs, and woody vines of Tennessee*. The University of Tennessee Press, Knoxville.

# APPENDIX I

# GLOSSARY

### A

**ACHENE** small, dry, one-locular, one-seeded indehiscent fruit with the seed coat and ovary wall separate.

**ACUMINATE** a tip whose sides are variously concave and tapering to a point.

**ACUTE** sharply ending in a point with margins straight or slightly convex.

**AERIAL** in the air, as in vines that attach to other objects with above-ground parts.

**AGGREGATE** a compound fruit composed of many ovaries from one flower, as in *Magnolia*.

**ARIL** an appendage or outer covering of the seed, typically fleshy and brightly colored, e.g., in *Celastrus*, *Euonymus*, and *Taxus*.

**AURICULATE** with auricles, or earlike lobes, at the base.

### B

**BERRY** a fleshy (pulpy or juicy) indehiscent fruit, with many seeds, these not enclosed by a hard endocarp; a berry from an inferior ovary has the calyx remnant attached at the apex of the fruit on the opposite side from the fruit stalk (as in *Ribes* and *Vaccinium*), whereas a berry from a superior ovary has the calyx remnant attached at the base of the fruit with the fruit stalk (as in *Diospyros* and *Vitis*); the "berry" of *Ilex* is actually a multi-seeded drupe, because the seeds are surrounded by a hard endocarp; many berry-named plants do not actually produce berries, e.g., blackberries (aggregate of drupes), chokeberries (pomes), hackberries (drupes), mulberries (multiple of drupes), and serviceberries (pomes).

**BRACT** a modified leaf associated with the inflorescence, usually attached at the base of pedicels or peduncles.

### C

**CALYX** collective term for all the sepals; the position of the calyx remnants on the fruit is helpful in determining the ovary position (see discussion under Berry).

**CAPSULE** a dry, dehiscent fruit derived from 2 or more carpels.

**CARPEL** the ancestral modified ovule-bearing leaf, often referred to as the basic unit of the ovary; unicarpellate fruits are 1-chambered, as in legumes and follicles, whereas multi-carpellate fruits are usually (but not always) multi-chambered.

**CARYOPSIS** fruit type of a grass, also referred to as a grain; single-seeded, the seed coat and fruit coat fused.

**CATKIN** (ament) a bracteate, often unisexual, apetalous, flexible spike-like or cymose inflorescence; the male inflorescence typically falling as a single unit.

**CILIATE** bearing marginal hairs, especially in reference to leaves and bracts.

**COMPOUND** divided into two or more similar parts.

**CONE** the seed-bearing structure of gymnosperms, in which the seeds are attached to an ovulate scale, and not enclosed in an ovary (as in angiosperms).

**CONICAL** cone-shaped.

**CONNATE** the fusion or joining together of similar structures.

**CORDATE** heart-shaped; with a sinus and rounded lobes, often in reference to the base of a structure.

**CORYMB** an indeterminate inflorescence, often broad and flattened, in which the central flower opens last.

**CRENATE** margins with shallow, round, or obtuse teeth.

**CRESCENT** a shape similar to the moon when it has one concave and one convex side illuminated.

**CUNEATE** wedge-shaped, straight-sided toward the base.

**CYLINDRIC** having the shape of a cylinder, especially a circular cylinder.

**CYME** a determinate inflorescence, often broad and flattened, in which the central flower opens first.

**CYPSELA** an achene derived from an inferior ovary, as in the Asteraceae, and often with a tuft of hairs at the apex.

### D

**DELTOID** shaped like a triangle.

**DENTATE** with sharp teeth pointed outward, at right angles to the midvein.

**Doubly Serrate** having coarse, upturned teeth, with smaller teeth along their margins.
**Drupe** a fleshy, indehiscent fruit usually with one seed enclosed in a hard endocarp.

## E
**Eglandular** without glands.
**Elliptic** oval; broadest near the middle and gradually tapering to both ends.

## F
**Follicle** a dry, unicarpellate fruit that splits along one side at maturity.

## G
**Glabrous** smooth, not hairy.
**Glandular** swollen-tipped hairs, often stalked, with the swollen portion bearing various types of secretions.
**Glaucous** covered with a whitish, waxy substance.
**Globose** round, globular, or spherical.

## H
**Head** dense cluster of flowers on an expanded receptacle, also called a capitulum, characteristic of the Asteraceae family; associated with a cypsela fruit type.
**Hip** the fleshy ripened hypanthium (floral cup) and enclosed achenes (fruits) of *Rosa*.
**Hirsute** with coarse or stiff hairs.

## I
**Inflorescence** the flower arrangement or mode of flower bearing.
**Internode** the portion of the stem between nodes (where leaf scars are located).

## K
**Keel** a prominent ridge.

## L
**Lanceolate** lance-shaped; narrow, and broadest near the base and tapering to the tip.
**Legume** a dry fruit from a single-carpellate ovary that splits along two lines or sutures, produced only by species in the bean family (Fabaceae).
**Linear** long and narrow with parallel sides throughout most of the length.
**Locule** compartment, cavity, or cell of an ovary, fruit, or anther.

## M
**Malodorous** having an unpleasant or offensive odor.
**Mucronate** with a short, abrupt tip, a mucro.
**Multiple** a compound fruit composed of many ovaries from many flowers, as in *Platanus*, *Liquidambar*, and *Morus*.

## N
**Node** the area of the twig where a leaf scar is located.
**Nut** a hard, dry, indehiscent, l-seeded fruit derived from an ovary with 2 or more carpels.
**Nutlet** a small nut, often associated with an involucre, as in *Carpinus*.

## O
**Oblanceolate** inversely lanceolate, i.e., narrow at the base and wider at the apex.
**Oblique** with an uneven base, asymmetrical.
**Oblong** longer than broad with more or less parallel sides.
**Obovate** inversely ovate, widest near the tip.
**Obtuse** blunt or rounded at the tip.
**Ovary** the portion of the pistil that bears the ovules; a mature ovary is a fruit and mature ovules are seeds; in a fruit derived from a superior ovary the calyx remnant is at the base of the fruit, whereas in a fruit derived from an inferior ovary, the calyx remnant is at the apex of the fruit.
**Ovate** egg-shaped, widest at the base.
**Ovoid** a solid that is oval in outline.

## P
**Palmate** lobed or divided in a palm- or hand-like manner.
**Panicle** an indeterminate branching raceme; the branches of the primary axis are racemose and the flowers are pedicellate.
**Pedicle** the stalk of an individual flower.
**Peduncle** the stalk supporting an inflorescence or the flower stalk of species producing solitary flowers.
**Peltate** umbrella- or shield-shaped, with the stalk attached centrally or within the margin.

**Perfoliate** leaf bases that completely surround the stem, the latter appearing to pass through the former.
**Pericarp** the fruit wall, sometimes divisible into an outer layer, the exocarp, a middle layer, the mesocarp, and an inner layer, the endocarp.
**Pinnate** with leaflets on both sides of a common axis.
**Plumose** resembling a feather; a central axis bearing fine hairs or side branches.
**Pome** a fleshy, indehiscent fruit derived from an inferior ovary and surrounded by an adnate hypanthium (e.g., pear, apple, and other related members of the Rosaceae); the calyx remnant is attached at the apex of the fruit, on the opposite side from the pedicel.
**Pseudocarp** a fibrous, persistent hypanthium enclosing several achenes, in this text restricted to *Calycanthus*.
**Pubescent** covered with soft hairs.
**Punctate** with translucent or colored dots, depressions, or pits.

## R

**Raceme** an elongate, unbranched, indeterminate inflorescence with pedicellate flowers.
**Receptacle** the tip of a pedicel or peduncle, usually enlarged, that bears the flower parts.
**Rhizome** an underground stem with nodes, buds, and roots.
**Rugose** covered with wrinkles.

## S

**Samara** an indehiscent, single-seeded, dry fruit with a prominent wing (e.g., *Acer*, *Fraxinus*, and *Ulmus*).
**Sepals** the usually leaflike flower structures (often 4 or 5 in number) attached either at the bottom of the fruit (in the case of a superior ovary, as in *Ilex* and *Diospyros*) or at the top of the fruit (in the case of an inferior ovary, as in *Vaccinium* or *Malus*).
**Serrate** a margin with sharp, forward-pointing teeth.
**Simple** undivided.
**Spikelet** the inflorescence of a grass, typically with a short axis bearing glumes and lemmas; associated with the caryopsis fruit type.
**Stellate** star-like; trichomes with radiating branches.
**Stipitate** with a stalk, often used to refer to a stalked, glandular hair.
**Stipule** an appendage, usually paired, at the base of a leaf petiole.
**Stolon** a horizontal stem, above or below ground level, that roots at the nodes and produces new plants at the stem tip.
**Style** the necklike portion of the ovary, often tipped by a knoblike stigma, sometimes branched and indicative of the number of locules in the ovary, often remnant as a stalk at the top of the fruit.
**Succulent** plants having thick, juicy leaves and/or stems.

## T

**Tendril** a modified leaf or stem by which a plant climbs or supports itself.
**Terete** round in cross-section.
**Tomentose** covered with short, dense, matted, and curly hairs.
**Truncate** square-tipped or square-based.

## U

**Umbel** a rounded or flat-topped flowering cluster with all the fruit stalks of the same length.

**APPENDIX II**

# WOODY PLANTS USEFUL FOR FOOD

Species in the following genera provide the most readily available sources of food in winter. Note: some frequently encountered herbaceous plants with green foliage and/or edible parts are also available in winter, including these non-native weeds: *Alliaria petiolata* (garlic mustard—leaves and stems), *Arctium minus* (burdock—roots), *Barbarea vulgaris* (winter cress—leaves and stems), *Stellaria media* (common chickweed—leaves and stems), and *Taraxacum officinale* (dandelion—roots, stems, and leaves).

Acer (maple)—sap
Aronia (chokeberries)—fleshy fruits
Betula (birch)—inner bark and sap
Carya (hickory)—nuts
Castanea (chestnut)—nuts
Corylus (hazelnut)—nuts
Crataegus (hawthorn)—fleshy fruits
Diospyros (persimmon)—fleshy fruits
Elaeagnus (autumn olive)—fleshy fruits
Fagus (beech)—nuts
Gleditsia (honey locust)—legumes
Juglans (walnut)—nuts
Juniperus (red-cedar)—fleshy cones
Malus/Pyrus (apple, crabapples)—fleshy fruits
Mitchella (partridge berry)—fleshy fruits
Opuntia (prickly pear)—fleshy pads and fruits
Pinus (pine)—inner bark
Platanus (sycamore)—sap
Quercus (oak)—nuts
Rhus (sumac)—fleshy fruits
Rosa (rose)—fleshy fruits
Smilax (greenbrier)—fleshy fruits
Sorbus (mountain-ash)—fleshy fruits
Tsuga (hemlock)—inner bark
Ulmus (elm)—inner bark
Viburnum (haw)—fleshy fruits

# WOODY PLANTS USEFUL FOR MEDICAL NEEDS

Species in the following genera are useful for certain emergency medical needs (such as healing of wounds) in winter. Further details are provided in Section IV. Note: some herbaceous plants with medicinal properties are also available in winter, including *Glechoma hederacea* (ground-ivy), *Stellaria media* (chickweed), and *Verbascum thapsus* (mullein).

Alnus (alder)
Hamamelis (witch-hazel)
Liquidambar (sweetgum)
Opuntia (prickly pear)
Pinus (pine)
Quercus (oak)

Salix (willow)
Tsuga (hemlock)
Ulmus (elm)
Vitis (grape)
Zanthoxylum (prickly-ash)

# Woody Plants Useful for Cordage

Species in these genera are good sources of cordage for making ropes, nets, and similar items. The bark of most trees is useful as cordage; it is best if fiber is taken from the dried inner bark of dead tree trunks or limbs for use as cordage; if living plants must be used, use the limbs, not the trunk, and if layers of fiber have to be removed from trunks, be careful not to girdle the tree. Among the best (indicated with ^) are basswood, willow, and slippery elm. Small roots of gymnosperms (*Juniperus*, *Picea*, and *Pinus*) are also a good source of cordage, as are the roots of *Juglans*, *Maclura*, and *Prunus*. Some herbaceous plants also provide good sources of fiber, including species of *Apocynum*, *Asclepias*, *Scirpus*, and *Typha*.

Acer (maple)
Asimina (pawpaw)
Broussonetia (paper mulberry)
Carya (hickory)
Corylus (hazelnut)
Fraxinus (ash)
Juglans (walnut)
Juniperus (red-cedar)
Maclura (osage-orange)
Morus (mulberry)
Picea (spruce)
Pinus (pine)
Populus (poplar, cottonwood)
Prunus (cherry, plum)
^Pueraria (kudzu)
Quercus (oak)
^Salix (willow)
Smilax (greenbrier)
^Tilia (basswood)
^Ulmus (elm)
Vitis (grape)

# Woody Plants Useful for Constructing Bows and Arrows

List of woody plants useful for making bows and arrows (an ^ indicates those woods considered to make the highest-quality arrows and bows).

**Arrows**
Amelanchier arborea/laevis (serviceberry)
Arundinaria gigantea (wild cane)
^Betula spp. (birch)
Cornus spp. (dogwood)
Ilex opaca (American holly)
^Quercus spp. (oak)
^Robinia pseudoacacia (black locust)
^Rosa spp. (rose)
^Salix spp. (willow)
^Ulmus spp. (elm)
Viburnum spp. (haw)

**Bows**
Arundinaria gigantea (wild cane)
Carya spp. (hickory)
Corylus americana (hazelnut)
Fraxinus americana (white ash)
Juglans nigra (black walnut)
^Juniperus virginiana (red-cedar)
^Maclura pomifera (osage-orange)
Malus spp. (apple, crabapple)
Pinus spp. (pine)
Quercus spp. (oak)
Robinia pseudoacacia (black locust)
^Taxus canadensis (yew)
Ulmus spp. (elm)

# INDEX OF SECTION I

Aerial roots, 14
Angiosperm, defined, 15
   Dicotyledonae (dicots), 17
   Monocotyledonae (monocots), 17
   vegetative features, 16
Annual growth rings, 16
Armature, corky ridges, and climbing adaptations
   aerial roots, 14
   bristle, 14
   corky ridges, 14
   prickle, 14
   spur shoot, 14
   stipular spine, 14
   tendrils, 14
   thorns, 14
Axillary buds, 13

Bark
   secondary phloem, 14
   secondary xylem, 14
   types of, 15
   vascular cambium, 14
Bows and arrows, 19
Branch scars 12,
Bristle, 14
Brown, Tom, Jr., 20
Buds
   axillary, 13
   catkin, 14
   collateral, 14
   false terminal, 13
   imbricate scaly, 13
   lateral, 13
   naked, 13
   scales, 13
   superposed, 14
   true terminal, 13
   valvate, 13
Bundle scars, 12, 13

Catkin, 14
Cedar-glade habitats, 16
Classification of gymnosperms and angiosperms, 17
Cones and fruits, 15
   dry and fleshy fruits, 15
Conifers, 17
Continuous pith, 14
Cordage, 19
Corky ridges, 14
Covered seed (angiosperm), 15

Deadly poisons, 19
Deciduous-leaved species, 4
Dicots, 17
Distributional abbreviations, 3

Edible parts, 20
Edible species, 19

FAC. *See* Facultative plants
FACU. *See* Facultative upland plants
FACW. *See* Facultative wetland plants
Facultative plants (FAC), 3
Facultative upland plants (FACU), 3
Facultative wetland plants (FACW), 3
False terminal bud, 13
   formation of, 13
Fibers, 18, 19
Food conducting cells, 16
Fruits scars, 12
Fruits and seed cones, 15, 17

Genera, numbers in Kentucky and Tennessee, 1
Geographic boundaries, 1, 2
Glabrous, 14
Growth rings, 14, 16
Gymnosperms vs. angiosperms, 15
   classifications of, 17
   cultivated ginkgos, 17
   ecological importance, 15
   definitions, 15
   history, 15
   notable facts, 16
   numbers of species, 15
   ovary, defined, 15
   reproductive differences, 17
   vegetative differences, 16

Hallucinations, 19
Herbaceous vs. woody plants, 1

Imbricate scales, 13
Inferior ovary, 15
Inflorescence types, 15
Inner bark, 14, 19
Internode, 12

Kentucky State Nature Preserves Commission, 17

Leaf, 12
   arrangements, 12
   descriptions, 12
   parts, 12
   planes of orientation, 12
   remnant, 12

Lenticels, 12

McPherson, Geri and John, 20
Medicinal uses, 19
Monocots, 17
Monocotyledonae.
   *See* Monocots

Naked buds, 13, 15
Naked seed (gymnosperm), 15
Native vs. non-native species, 1
NI. *See* No indicator
Node, 12
No indicator (NI), 3
Non-native species, 1, 3, 18, 20
Non-native woody plants
   ecological threats, 18
   edible, 20
   future threats, 18
   shrubs, 18
   trees, 18
   vines, 18
Numbers of woody genera and species in Kentucky and Tennessee, 1

OBL. *See* Obligate wetland plant
Obligate wetland plant (OBL), 3
Ovary, defined, 15
   Inferior vs. superior, 15

Petiole, 12, 13
Petiolule, 12
Phloem, 13, 14, 16
Physiographic provinces, 2, 3
Pistil, 15
Pith, 6, 14
Planes of leaf orientation, 12, 13
Plants,
   edibility, 18, 19
   medicinal use, 19
   poisonous look-alikes, 19
   rope and string, 19
   self-medication, 19, 20
   types to avoid as food, 19
   weapons, 19
Poisonous plants, 19
Pollination, 17
Prickle, 14
Protective scales, 13
Pubescence, 14
   glabrous, 14
   stellate, 14

Rare woody plants, 17
   list of in Tennessee and Kentucky, 17–18
Remnant fruits, 4, 17
Reproductive differences between gymnosperms and angiosperms, 17

Scales, 13, 15, 16
Scars, twig, 12
   bud scale scar, 13
   branch scar, 12
   fruit scar, 12
   leaf scar, 13
   stipule scar, 13
   vascular bundle scar, 13
Secondary phloem, 14
Secondary xylem, 14
Seed, 15, 17
Seed production, 17
Shrubs, defined, 1,
Species,
   how to identify using the keys, 2
   importance of proper identification, 18
   numbers in Kentucky and Tennessee, 1
Spikelets, 17
Spring wood cells, 16
Spur shoot, 14
Standard two-letter abbreviations, 3
Stellate, 14

Stems
   dicots vs. monocots, 17
Sterile perianth parts, 15
Stipular spine, 14
Stipules, 13, 14
Subtropical U.S. regions, cycads present, 17
Superior ovary, 15

Tendrils, 14
Tennessee Natural Heritage Program, 17
Thorn
   with buds, 14
   lacking buds, 14
Trees, defined, 1
Trichomes, 14
Trunks, avoidance of girdling, 19
Twigs,
   internodes, 12
   lenticels, 12
   nodes, 12

UPL. *See* Upland plants
Universal Edibility Test, 19
Upland plants (UPL), 3

Valvate, 13
Vascular bundle scars, 13
Vascular bundles, 16, 17
Vascular cambium, 14
Vascular tissue, 14
Vegetative differences between gymnosperms and angiosperms features, 16

   leaf types and production, 16
   growth rings in, 16
   porous and nonporous, 16
   vascular systems, 16
   wood of, 16

Water-conducting cells, 16
Weapons, 19, 20
Wetland indicator categories, 3
   Facultative plants (FAC), 3
   Facultative upland plants (FACU), 3
   Facultative wetland plants (FACW), 3
   Obligate wetland plant (OBL), 3
   No indicator (NI), 3
   Upland plants (UPL), 3
Wilderness survival skills, 20
Winter emergencies, 19
Winter identification, 4, 13
Wood selection, 20
Woody plants
   conservation concerns, 20
   definitions, 1
   as food, medicine, fiber, and weapons, 18–20
   non-native, 18
   plants to avoid when using for food, 20
   prickle, 14
   rare types of, 17
Woody vines, defined, 1

Xylem, 13

# INDEX OF SCIENTIFIC NAMES IN SECTIONS IV AND V

Entries in **boldface** are Plate numbers, which can be found in Section V.

Abies, 45
  fraseri, 45, **1**
Acer, 45
  drummondii, 46, **2**
  floridanum, 46
  leucoderme, 46
  negundo, 46, **3**, **4**
  nigrum, 46
  pensylvanicum, 46, **5**, **6**
  platanoides, 46, **7**
  rubrum var. rubrum, 46, **8**
  rubrum var. trilobum, 46
  saccharinum, 46, **9**
  saccharum, 46, **10**, **11**
  spicatum, 46, **12**
Aesculus, 46
  flava, 47, **13**, **14**
  glabra, 47, **15**, **16**
  pavia, 47, **17**
  sylvatica, 47
Ailanthus, 47
  altissima, 47, **18**, **19**
Akebia, 47
  quinata, 47, **20**
Albizia, 47
  julibrissin, 48, **21**, **22**
Alnus, 48
  glutinosa, 48
  serrulata, 48, **23**, **24**, **25**
  viridis subsp. crispa, 48, **26**
Amelanchier, 48
  arborea, 48, **27**
  canadensis, 49
  laevis, 49
  sanguinea, 49
Amorpha, 49
  fruticosa, 49, **28**, **29**
  glabra, 49
  nitens, 49, **30**, **31**
Ampelopsis, 49
  arborea, 49, **32**, **33**
  brevipedunculata, 49
  cordata, 49, **34**
Aralia, 49
  spinosa, 50, **35**, **36**
Aristolochia, 50
  macrophylla, 50, **37**
  tomentosa, 50, **38**, **39**

Aronia, 50
  arbutifolia, 50, **40**
  melanocarpa, 50, **41**, **42**
Arundinaria, 50
  appalachiana, 51, **43**
  gigantea, 51, **44**
Asimina, 51
  triloba, 51, **45**
Baccharis, 51
  halimifolia, 51, **46**, **47**
Berberis, 51
  *bealei*, 80
  canadensis, 51, **48**
  thunbergii, 51, **49**
Berchemia, 51
  scandens, 52, **50**, **51**
Betula, 52
  alleghaniensis, 52, **52**, **53**, **54**, **55**
  cordifolia, 52, **56**
  lenta, 52, **57**, **58**, **59**
  nigra, 52, **60**, **61**
Bignonia, 52
  capreolata, 52, **62**, **63**, **64**
Broussonetia, 53
  papyrifera, 53, **65**, **66**
Brunnichia, 53
  ovata, 53, **67**, **68**
Buckleya, 53
  distichophylla, 53, **69**
Buddleja, 53
  davidii, 53, **70**
*Bumelia*, 103
  *lycioides*, 103
*Buxella*, 69
  *brachycera*, 69

Callicarpa, 53
  americana, 53, **71**, **72**
Calycanthus, 54
  floridus var. floridus, 54, **73**, **74**
  floridus var. glaucus, 54
Calycocarpum, 54
  lyonii, 54, **75**, **76**
Campsis, 54
  radicans, 54, **77**, **78**
Carpinus, 54
  caroliniana, 54, **79**
Carya, 54
  *alba*, 56

  aquatica, 55, **80**, **81**
  *carolinae-septentrionalis*, 55
  cordiformis, 55, **82**, **83**
  glabra, 55, **84**, **85**
  illinoinensis, 55, **86**, **87**, **88**
  laciniosa, 55, **89**, **90**
  *ovalis*, 55
  ovata var. australis, 55, **91**
  ovata var. ovata, 55, **92**, **93**
  pallida, 55, **94**, **95**
  tomentosa, 56, **96**, **97**
Castanea, 56
  dentata, 56, **98**, **99**, **100**
  mollissima, 56, **101**, **102**
  pumila, 56, **103**, **104**
Catalpa, 56
  bignonioides, 56, **105**, **106**
  speciosa, 56, **107**, **108**
Ceanothus, 57
  americanus, 57, **109**, **110**
  herbaceus, 57
Celastrus, 57
  orbiculatus, 57, **111**, **112**
  scandens, 57, **113**
Celtis, 57
  laevigata, 57
  occidentalis, 58, **114**
  tenuifolia, 58, **115**
Cephalanthus, 58
  occidentalis, 58, **116**
Cercis, 58
  canadensis, 58, **117**, **118**
Chaenomeles, 58
  speciosa, 58, **119**
Chimaphila, 58
  maculata, 58, **120**
Chionanthus, 58
  virginicus, 59, **121**
Cladrastis, 59
  kentukea, 59, **122**, **123**
Clematis, 59
  catesbyana, 59, **124**
  *dioscoreifolia*, 59
  *paniculata*, 59
  terniflora, 59, **125**
  virginiana, 59, **126**, **127**
Clethra, 59
  acuminata, 60, **128**, **129**
  alnifolia, 60, **130**

Cocculus, 60
　carolinus, 60, **131**, **132**
Comptonia, 60
　peregrina, 60, **133**, **134**
Conradina, 60
　verticillata, 60, **135**
Cornus, 60
　alternifolia, 61, **136**
　amomum subsp. amomum, 61, **137**
　amomum subsp. obliqua, 61
　drummondii, 61, **138**
　florida, 61, **139**, **140**
　foemina subsp. foemina, 61, **141**, **142**
　foemina subsp. racemosa, 61
　sericea, 61
　*stricta*, 61
Corylus, 61
　americana, 61, **143**, **144**, **145**
　cornuta, 61, **146**, **147**, **148**
Cotinus, 61
　obovatus, 62, **149**, **150**
Crataegus, 62
　berberifolia, 62
　*boyntonii*, 62
　calpodendron, 62
　coccinea, 62
　collina, 62
　crus-galli, 62
　disperma, 62
　*flabellata*, 62
　harbisonii, 62, **151**
　intricata, 62
　macrosperma, 62
　marshallii, 62
　mollis, 62
　phaenopyrum, 62, **152**
　pruinosa, 63
　punctata, 63
　*rubella*, 62
　spathulata, 63
　uniflora, 63
　viridis, 63, **153**
Cytisus, 63
　scoparius, 63, **154**

Decodon, 63
　verticillatus, 63, **155**
Decumaria, 63
　barbara, 63, **156**, **157**
Deutzia, 63
　scabra, 64, **158**, **159**
Diervilla, 64
　lonicera, 64, **160**
　rivularis, 64, **161**, **162**
　sessilifolia, 64, **163**, **164**
Diospyros, 64
　virginiana, 64, **165**, **166**
Dirca, 64
　palustris, 64, **167**

Elaeagnus, 64–65
　angustifolia, 65
　multiflora, 65
　pungens, 65, **168**, **169**
　umbellata, 65, **170**
Epigaea, 65
　repens, 65, **171**, **172**
*Eubotrys*, 76
　*racemosa*, 76
　*recurva*, 76
Euonymus, 65
　alatus, 66, **173**, **174**
　americanus, 66, **175**, **176**
　atropurpureus, 66, **177**, **178**
　europaeus, 66
　fortunei, 66, **179**
　*hederaceus*, 66
　japonicus, 66
　kiautschovicus, 66
　obovatus, 66, **180**, **181**

Fagus, 66
　grandifolia, 66, **182**, **183**, **184**
Forestiera, 66-67
　acuminata, 67, **185**
　ligustrina, 67, **186**
Forsythia, 67
　suspensa, 67
　viridissima, 67, **187**
Fothergilla, 67
　major, 67, **188**, **189**
Frangula, 67
　alnus, 67
　caroliniana, 67, **190**
Fraxinus, 67
　americana, 68, **191**, **192**
　pennsylvanica var. pennsylvanica, 68, **193**, **194**
　pennsylvanica var. subintegerrima, 68
　profunda, 68, **195**, **196**
　quadrangulata, 68, **197**, **198**

Gaultheria, 68
　procumbens, 68, **199**
Gaylussacia, 68
　baccata, 69, **200**
　brachycera, 69, **201**
　dumosa, 69, **202**
　ursina, 69, **203**
Gelsemium, 69
　sempervirens, 69, **204**, **205**
Gleditsia, 69
　aquatica, 69, **206**, **207**
　triacanthos, 69, **208**, **209**
Gymnocladus, 69–70
　dioicus, 70, **210**, **211**, **212**

Halesia, 70
　carolina, 70, **213**, **214**, **215**
　*tetraptera*, 70
Hamamelis, 70
　virginiana, 70, **216**, **217**

Hedera, 70
　helix, 70, **218**, **219**
Hibiscus, 70–71
　syriacus, 71, **220**, **221**
Hydrangea, 71
　arborescens, 71, **222**, **223**
　cinerea, 71, **224**, **225**
　quercifolia, 71, **226**, **227**
　radiata, 71, **228**
Hypericum, 71
　crux-andreae, 72, **229**, **230**
　densiflorum, 72, **231**
　frondosum, 72, **232**
　hypericoides, 72, **233**
　lobocarpum, 72, **234**, **235**
　nudiflorum, 72, **236**
　prolificum, 72, **237**, **238**
　stragulum, 72, **239**

Ilex, 72
　ambigua var. ambigua, 73
　ambigua var. montana, 73, **240**
　*collina*, 82
　decidua, 73, **241**, **242**
　longipes, 73, **243**
　opaca, 73, **244**
　verticillata, 73, **245**, **246**, **247**
*Isotrema*, 50
　*macrophylla*, 50
　*tomentosa*, 50
Itea, 73
　virginica, 73, **248**, **249**, **250**
Juglans, 73–74
　cinerea, 74, **251**, **252**
　nigra, 74, **253**, **254**
Juniperus, 74
　communis, 74
　virginiana, 74, **255**, **256**
Kalmia, 74
　*buxifolia*, 75
　carolina, 74, **257**
　latifolia, 74, **258**, **259**
Kerria, 74–75
　japonica, 75, **260**, **261**
Koelreuteria, 75
　paniculata, 75, **262**, **263**
Leiophyllum, 75
　buxifolium, 75, **264**
Lespedeza, 75
　bicolor, 75, **265**, **266**
　thunbergii, 75
Leucothoe, 75
　fontanesiana, 76, **267**, **268**, **269**
　racemosa, 76, **270**
　recurva, 76, **271**, **272**
Ligustrum, 76
　amurense, 76
　japonicum, 76
　obtusifolium, 76
　ovalifolium, 76

## INDEX OF SCIENTIFIC NAMES IN SECTIONS IV AND V

sinense, 76, **273**, **274**, **275**
vulgare, 76
Lindera, 76
  benzoin, 76, **276**, **277**
Linnaea, 76–77
  borealis, 77, **278**
Liquidambar, 77
  styraciflua, 77, **279**, **280**
Liriodendron, 77
  tulipifera, 77, **281**, **282**, **283**
Lonicera, 77
  canadensis, 78
  dioica var. diocia, 78, **284**
  dioica var. glaucescens, 78
  dioica var. orientalis, 78
  flava, 78, **285**
  fragrantissima, 78, **286**, **287**
  japonica, 78, **288**, **289**, **290**, **291**
  maackii, 78, **292**, **293**, **294**
  morrowii, 78
  *prolifera*, 78
  reticulata, 78
  sempervirens, 78, **295**
Lycium, 78
  barbarum, 78, **296**
Lyonia, 79
  ligustrina var. foliosiflora, 79
  ligustrina var. ligustrina, 79, **297**, **298**

Maclura, 79
  pomifera, 79, **299**, **300**, **301**
Magnolia, 79
  acuminata, 80, **302**
  fraseri, 80, **303**, **304**
  grandiflora, 80, **305**
  macrophylla, 80, **306**, **307**, **308**
  pyramidata, 80
  tripetala, 80, **309**, **310**
  virginiana, 80, **311**
Mahonia, 80
  bealei, 80, **312**
Malus, 80
  angustifolia, 80, **313**
  coronaria, 80, **314**
  ioensis, 81
  pumila, 81, **315**
Melia, 81
  azedarach, 81, **316**, **317**
Menispermum, 81
  canadense, 81, **318**, **319**
Menziesia, 81
  pilosa, 81, **320**, **321**
Mitchella, 81
  repens, 81, **322**
Morus, 81–82
  alba, 82, **323**
  *murrayana*, 82
  rubra, 82, **324**

Nandina, 82
  domestica, 82, **325**, **326**
Nemopanthus, 82
  collinus, 82, **327**
Nestronia, 82
  umbellula, 82, **328**, **329**
Neviusia, 82
  alabamensis, 83, **330**, **331**
Nyssa, 83
  aquatica, 83, **332**
  biflora, 83, **333**
  sylvatica, 83, **334**, **335**, **336**

Opuntia, 83
  *compressa*, 83
  humifusa, 83, **337**
Ostrya, 83
  virginiana, 83, **338**, **339**
Oxydendrum, 83–84
  arboreum, 84, **340**, **341**

Parthenocissus, 84
  quinquefolia, 84, **342**
Paulownia, 84
  tomentosa, 84, **343**, **344**, **345**, **346**
Paxistima, 84
  canbyi, 84, **347**
Philadelphus, 84
  hirsutus, 85, **348**
  inodorus, 85, **349**, **350**
  pubescens, 85, **351**
Phoradendron, 85
  leucarpum subsp. leucarpum, 85, **352**
*Photinia*, 50
  *floribunda*, 50
  *melanocarpa*, 50
  *prunifolia*, 50
  *pyrifolia*, 50
Physocarpus, 85
  opulifolius, 85, **353**, **354**, **355**
Picea, 85
  rubens, 85, **356**
Pieris, 85–86
  floribunda, 86, **357**
Pinus, 86
  echinata, 86, **358**
  pungens, 86, **359**
  rigida, 86, **360**
  strobus, 86, **361**
  taeda, 86, **362**
  virginiana, 86, **363**
Planera, 87
  aquatica, 87, **364**
Platanus, 87
  occidentalis, 87, **365**, **366**
Polygonella, 87
  americana, 87, **367**, **368**

Poncirus, 87
  trifoliata, 87, **369**
Populus, 87–88
  alba, 88, **370**
  balsamifera, 88
  deltoides, 88, **371**
  grandidentata, 88, **372**
  heterophylla, 88, **373**
  × jackii, 88, **374**
*Potentilla*, 103
  *tridentata*, 103
Prunus, 88
  americana, 89, **375**
  angustifolia, 89, **376**
  avium, 89
  cerasus, 89, **377**
  hortulana, 89
  mahaleb, 89, **378**
  mexicana, 89
  munsoniana, 89
  pensylvanica, 89, **379**
  persica, 89, **380**
  pumila, 89, **381**
  serotina, 89, **382**
  umbellata, 90, **383**
  virginiana, 90, **384**
Ptelea, 90
  trifoliata, 90, **385**
Pueraria, 90
  montana var. lobata, 90, **386**, **387**
Pyrularia, 90
  pubera, 90, **388**
Pyrus, 90–91
  calleryana, 91, **389**, **390**, **391**
  communis, 91, **392**

Quercus, 91
  alba, 93, **393**, **394**
  bicolor, 93, **395**, **396**
  coccinea, 93, **397**, **398**
  falcata, 93, **399**, **400**
  ilicifolia, 93
  imbricaria, 93, **401**, **402**
  lyrata, 93, **403**, **404**
  macrocarpa, 93, **405**, **406**
  margaretta, 93, **407**
  marilandica, 94, **408**, **409**
  michauxii, 94, **410**, **411**
  montana, 94, **412**, **413**
  muhlenbergii, 94, **414**, **415**
  nigra, 94, **416**, **417**
  *nuttallii*, 94
  pagoda, 94, **418**, **419**
  palustris, 94, **420**, **421**
  phellos, 94, **422**, **423**
  prinoides, 94, **424**
  *prinus*, 94
  rubra, 94, **425**, **426**

shumardii, 94, **427**, **428**
stellata, 94, **429**, **430**
texana, 94, **431**, **432**
velutina, 94, **433**, **434**

Rhamnus, 94–95
alnifolia, 95, **435**
*caroliniana*, 67
cathartica, 95
*citrifolia*, 95
davurica, 95, **436**
lanceolata, 95, **437**
Rhododendron, 95
alabamense, 95, **438**, **439**
arborescens, 96, **440**, **441**
*bakeri*, 96
calendulaceum, 96, **442**, **443**
canescens, 96, **444**, **445**
catawbiense, 96, **446**, **447**
cumberlandense, 96, **448**, **449**
maximum, 96, **450**, **451**
minus, 96, **452**
periclymenoides, 96, **453**, **454**
prinophyllum, 96
viscosum, 96, **455**
Rhodotypos, 96
scandens, 96, **456**, **457**
Rhus, 96
aromatica, 97, **458**
copallinum, 97, **459**
glabra, 97, **460**, **461**
*radicans*, 108
*toxicodendron*, 108
typhina, 97, **462**, **463**
*vernix*, 108
Ribes, 97
americanum, 97
aureum var. villosum, 97, **464**
curvatum, 97, **465**, **466**
cynosbati, 98, **467**
glandulosum, 98, **468**
missouriense, 98, **469**
*odoratum*, 98
rotundifolium, 98, **470**
Robinia, 98
hispida, 98, **471**
pseudoacacia, 98, **472**, **473**, **474**
Rosa, 98
carolina, 99, **475**, **476**
eglanteria, 99
multiflora, 99, **477**, **478**
palustris, 99, **479**, **480**
setigera, 99, **481**, **482**
virginiana, 99, **483**, **484**
wichuraiana, 99
Rubus, 99
allegheniensis, 100, **485**
argutus, 100, **486**, **487**
*baileyanus*, 100
*betulifolius*, 100

bifrons, 100, **488**, **489**
canadensis, 100, **490**
*deamii*, 100
*enslenii*, 100
*depavitus*, 100
*fecundus*, 100
*felix*, 100
flagellaris, 100, **491**
hispidus, 100, **492**
idaeus subsp. strigosus, 100, **493**, **494**
*indianensis*, 100
*invisus*, 100
*kentuckiensis*, 100
*levicaulis*, 100
longii, 100, **495**, **496**
*meracus*, 100
occidentalis, 101, **497**
odoratus, 101, **498**
pensilvanicus, 101
phoenicolasius, 101, **499**
*roribaccus*, 100
trivialis, 101, **500**
*whartoniae*, 100

Salix, 101
alba, 102, **501**
amygdaloides, 101
babylonica, 102, **502**, **503**
caroliniana, 102, **504**
cinerea, 102
discolor, 101
eriocephala, 102, **505**
exigua, 102, **506**
fragilis, 101
humilis var. humilis, 102, **507**
humilis var. microphylla, 102
nigra, 102, **508**
pentandra, 101
purpurea, 101
sericea, 102, **509**
Sambucus, 102
canadensis, 102, **510**, **511**
racemosa subsp. pubens, 102, **512**, **513**
Sassafras, 102-103
albidum, 103, **514**
Schisandra, 103
glabra, 103, **515**
Sibbaldiopsis, 103
tridentata, 103, **516**
Sideroxylon, 103
lycioides, 103, **517**, **518**
Smilax, 103
bona-nox, 104, **519**, **520**
glauca, 104, **521**, **522**
*hispida*, 104
laurifolia, 104, **523**, **524**
rotundifolia, 104, **525**, **526**, **527**
tamnoides, 104, **528**
walteri, 104, **529**

Solanum, 104
dulcamara, 104, **530**, **531**
Sorbus, 104–105
americana, 105, **532**, **533**
Spiraea, 105
alba, 105, **534**, **535**
japonica, 105, **536**, **537**
prunifolia, 105, **538**
tomentosa, 105, **539**, **540**
virginiana, 105, **541**
Staphylea, 105
trifolia, 105, **542**, **543**
Stewartia, 106
ovata, 106, **544**
Styrax, 106
americanus, 106, **545**
grandifolius, 106, **546**
Symphoricarpos, 106
albus, 106
orbiculatus, 106, **547**, **548**
Symplocos, 106
tinctoria, 107, **549**, **550**, **551**
Syringa, 107
vulgaris, 107, **552**

Taxodium, 107
distichum, 107, **553**, **554**
Taxus, 107
baccata, 107
canadensis, 107, **555**, **556**
cuspidata, 107
Thuja, 107
occidentalis, 108, **557**
Tilia, 108
americana var. americana, 108
americana var. heterophylla, 108, **558**
Toxicodendron, 108
pubescens, 108, **559**, **560**
radicans, 108, **561**, **562**
vernix, 108, **563**, **564**, **565**
Trachelospermum, 108
difforme, 109, **566**, **567**
Tsuga, 109
canadensis, 109, **568**, **569**
caroliniana, 109, **570**

Ulmus, 109
alata, 109, **571**, **572**
americana, 109, **573**
crassifolia, 109, **574**
rubra, 110, **575**
serotina, 110, **576**
thomasii, 110, **577**, **578**

Vaccinium, 110
arboreum, 111, **579**
corymbosum, 111, **580**, **581**, **582**
elliottii, 111, **583**
erythrocarpum, 111, **584**
fuscatum, 111, **585**
hirsutum, 111, **586**

macrocarpon, 111, **587**, **588**
pallidum, 111, **589**
stamineum var.
　melanocarpum, 111, **590**
stamineum var. stamineum, 111
Viburnum, 111
　acerifolium, 112, **591**
　*alnifolium*, 112
　bracteatum, 112, **592**
　cassinoides, 112, **593**
　dentatum, 112, **594**
　lantanoides, 112, **595**
　molle, 112, **596**, **597**
　nudum, 112, **598**
　*nudum* var. *cassinoides*, 112
　prunifolium, 112, **599**, **600**
　rafinesquianum var. affine, 112
　rafinesquianum var.
　　rafinesquianum, 112, **601**
　rhytidophyllum, 112, **602**
　rufidulum, 113, **603**, **604**
Vinca, 113
　major, 113, **605**
　minor, 113, **606**
Vitis, 113
　aestivalis, 113, **607**, **608**
　cinerea var. baileyana, 114, **609**, **610**, **611**
　cinerea var. cinerea, 114
　labrusca, 114, **612**
　labruscana, 114
　palmata, 114, **613**
　riparia, 114, **614**, **615**
　rotundifolia, 114, **616**
　rupestris, 114, **617**, **618**
　vulpina, 114, **619**, **620**, **621**
Wisteria, 114
　floribunda, 114, **622**, **623**
　frutescens, 114, **624**, **625**
　sinensis, 114
Xanthorhiza, 114–115
　simplicissima, 115, **626**
Yucca, 115
　filamentosa, 115, **627**, **628**
　flaccida, 115, **629**
Zanthoxylum, 115
　americanum, 115, **630**

# INDEX OF COMMON NAMES IN SECTION IV

Common names are based on Jones (2005) and Chester et al. (2009). For unhyphenated common names, the names are alphabetized by the last name, followed by the first and second name. For example, the common roundleaf greenbrier is listed as: "greenbrier, common roundleaf." In the case of a hyphenated name, such as eastern red-cedar and swamp-privet (the names are hyphenated in these cases because they are not a true cedar or privet, respectively), the names are listed as follows: "red-cedar, eastern," and "swamp-privet."

Akebia, five-leaf, 47
Alder, black, 48
Alder, mountain, 48
Alder, smooth, 48
Allspice, Carolina, 54
Apple, common, 81
Arborvitae, 108
Arbutus, trailing, 65
Arrow-wood, 112
Arrow-wood, limerock, 112
Ash, blue, 68
Ash, green, 68
Ash, pumpkin, 68
Ash, white, 68
Asian-olive, 65
Aspen, bigtooth, 88
Autumn-olive, 65
Azalea, Alabama, 95
Azalea, Cumberland, 96
Azalea, flame, 96
Azalea, pink, 96
Azalea, rosebud, 96
Azalea, smooth, 96
Azalea, southern pinxter, 96
Azalea, swamp, 96
Azalea, sweet, 96

Bald-cypress, 107
Balm-of-Gilead, 88
Barberry, Canada, 51
Barberry, Japanese, 51
Basswood, 108
Bayonet, Spanish, 115
Bearberry, 111
Beautyberry, American, 53
Beech, American, 66
Bells, golden, 67
Birch, heart-leaf, 52
Birch, river, 52
Birch, sweet, 52
Birch, yellow, 52
Bittersweet, American, 57
Bittersweet, oriental, 57
Blackberry, Allegheny, 100
Blackberry, Himalayan, 100
Blackberry, Long's, 100

Blackberry, Pennsylvania, 101
Blackberry, smooth, 100
Blackberry, southern, 100
Blackgum, 83
Bladder-nut, 105
Blueberry, black high-bush, 111
Blueberry, hairy, 111
Blueberry, high-bush, 111
Blueberry, low-bush, 111
Bodark, 79
Bodock, 79
Boxelder, 46
Broom, Scotch, 63
Buckbush, 73
Buckeye, Ohio, 47
Buckeye, painted, 47
Buckeye, red, 47
Buckeye, scarlet, 47
Buckeye, sweet, 47
Buckeye, yellow, 47
Buckthorn, alderleaf, 95
Buckthorn, Carolina, 67
Buckthorn, Dahurian, 95
Buckthorn, European, 95
Buckthorn, lanceleaf, 95
Buckthorn, southern, 103
Buffalo-nut, 90
Burning-bush, winged, 66
Butterfly-bush, 53
Buttonbush, common, 58

Camellia, mountain, 106
Cane, giant, 51
Cane, hill, 51
Catalpa, northern, 56
Catalpa, southern, 56
Cherry, black, 89
Cherry, fire, 89
Cherry, Mahaleb, 89
Cherry, sand, 89
Cherry, sour, 89
Cherry, sweet, 89
Chestnut, American, 56
Chestnut, Chinese, 56
Chestnut, dwarf, 56
Chinaberry, 81

Chinkapin, Allegheny, 56
Chokeberry, black, 50
Chokeberry, red, 50
Chokecherry, 90
Clematis, sweet autumn, 59
Clover, bush, 75
Coffeetree, Kentucky, 70
Conjurer's-nut, 82
Coralbeads, 60
Coralberry, 106
Cottonwood, eastern, 88
Cottonwood, swamp, 88
Crabapple, Iowa, 81
Crabapple, southern, 80
Crabapple, sweet, 80
Cranberry, 111
Creeper, trumpet, 54
Creeper, winter, 66
Crossvine, 52
Cupseed, Lyon's, 54
Currant, buffalo, 97
Currant, skunk, 98
Currant, wild black, 97

Deerberry, 111
Dewberry, Coastal Plain, 101
Dewberry, northern, 100
Dewberry, swamp, 100
Dogbane, climbing, 109
Dog-hobble, highland, 76
Dog-hobble, redtwig, 76
Dog-hobble, swamp, 76
Dogwood, alternate-leaved, 61
Dogwood, flowering, 61
Dogwood, red osier, 61
Dogwood, rough-leaved, 61
Dogwood, silky, 61
Dogwood, stiff, 61

Elderberry, common, 102
Elderberry, red, 102
Elm, American, 109
Elm, cedar, 109
Elm, red, 110
Elm, rock, 110
Elm, September, 110

# INDEX OF COMMON NAMES IN SECTION IV

Elm, slippery, 110
Elm, winged, 109
Euonymus, climbing, 66

Fetterbush, 76
Fetterbush, mountain, 86
Fir, Fraser, 45
Fivefingers, dwarf, 103
French-mulberry, 53
Fringetree, 59

Gooseberry, Appalachian, 98
Gooseberry, eastern prickly, 98
Gooseberry, granite, 97
Gooseberry, Missouri, 98
Grape, cat, 114
Grape, downy, 114
Grape, fox, 114
Grape, frost, 114
Grape, graybark, 114
Grape, muscadine, 114
Grape, red, 114
Grape, riverside, 114
Grape, sand, 114
Grape, summer, 113
Grape-holly, Oregon, 80
Greenbrier, bristly, 104
Greenbrier, cat, 104
Greenbrier, common roundleaf, 104
Greenbrier, laurel, 104
Greenbrier, red-berried, 104
Greenbrier, saw, 104

Hackberry, dwarf, 58
Hackberry, northern, 58
Hardhack, 105
Haw, possum, 73, 112
Haw, rusty black, 113
Hawthorn, barberry, 62
Hawthorn, cockspur, 62
Hawthorn, dotted, 63
Hawthorn, downy, 62
Hawthorn, entangled, 62
Hawthorn, fan-leaf, 62
Hawthorn, frosted, 63
Hawthorn, green, 63
Hawthorn, Harbison's, 62
Hawthorn, hill, 62
Hawthorn, littlehip, 63
Hawthorn, one-flower, 63
Hawthorn, parsley, 62
Hawthorn, pear, 62
Hawthorn, scarlet, 62
Hawthorn, spreading, 62
Hawthorn, Washington, 62
Hazelnut, American, 61
Hazelnut, beaked, 61
Hearts-a-bursting-with-love, 66
Heavenly-bamboo, 82
Hedge-apple, 79
Hemlock, Carolina, 109

Hemlock, eastern, 109
Hickory, bitternut, 55
Hickory, mockernut, 56
Hickory, pecan, 55
Hickory, pignut, 55
Hickory, red, 55
Hickory, sand, 55
Hickory, shagbark, 55
Hickory, shellbark, 55
Hickory, water, 55
Hickory, white, 56
Hobble, witch, 112
Holly, American, 73
Holly, Appalachian mountain, 82
Holly, Carolina, 73
Honey-locust, 69
Honeysuckle, Amur, 78
Honeysuckle, fly, 78
Honeysuckle, grape, 78
Honeysuckle, Japanese, 78
Honeysuckle, Morrow's, 78
Honeysuckle, mountain bush, 64
Honeysuckle, northern bush, 64
Honeysuckle, southern bush, 64
Honeysuckle, trumpet, 78
Honeysuckle, wild, 78
Honeysuckle, yellow, 78
Hophornbeam, 83
Hop-tree, common, 90
Hornbeam, American, 54
Horse-apple, 79
Huckleberry, bear, 69
Huckleberry, black, 69
Huckleberry, box, 69
Huckleberry, dwarf, 69
Hydrangea, climbing, 63
Hydrangea, gray, 71
Hydrangea, oak-leaved, 71
Hydrangea, silverleaf, 71
Hydrangea, wild, 71

Indigo, mountain false, 49
Indigo, shining false, 49
Indigo, tall false, 49
Ironwood, 54, 83
Ivy, English, 70

Jessamine, yellow, 69
Jetbead, 96
Jointweed, southern, 87
Juniper, common, 74

Kudzu, 90

Laurel, Carolina, 74
Laurel, great, 96
Laurel, mountain, 74
Laurel, rosebay, 96
Leatherwood, 64
Lespedeza, shrub, 75
Lilac, common, 107

Locust, black, 98
Locust, bristly, 98
Loosestrife, swamp, 63

Magnolia, bigleaf, 80
Magnolia, cucumber, 80
Magnolia, Fraser, 80
Magnolia, southern, 80
Magnolia, umbrella, 80
Maleberry, 79
Maple, black, 46
Maple, chalk, 46
Maple, Drummond red, 46
Maple, Florida, 46
Maple, mountain, 46
Maple, Norway, 46
Maple, red, 46
Maple, silver, 46
Maple, striped, 46
Maple, sugar, 46
Maple, water, 46
Matrimony-vine, 78
Mayberry, 111
Meadowsweet, white, 105
Mimosa, 48
Minnie-bush, 81
Mistletoe, 85
Mock-orange, 64
Mock-orange, Appalachian, 85
Mock-orange, Cumberland, 85
Mock-orange, Ozark, 85
Moonseed, common, 81
Mountain-ash, American, 105
Mountain-lover, Canby's, 84
Mulberry, red, 82
Mulberry, white, 82

Needle, Adam's, 115
Nightshade, climbing, 104
Ninebark, common, 85

Oak, bear, 93
Oak, black, 94
Oak, blackjack, 94
Oak, bur, 93
Oak, cherrybark, 94
Oak, chestnut, 94
Oak, chinkapin, 94
Oak, dwarf chinkapin, 94
Oak, dwarf post, 93
Oak, northern red, 94
Oak, Nuttall's, 94
Oak, overcup, 93
Oak, pin, 94
Oak, post, 94
Oak, scarlet, 93
Oak, shingle, 93
Oak, Shumard, 94
Oak, southern red, 93

Oak, Spanish, 93
Oak, swamp chestnut, 94
Oak, swamp white, 93
Oak, Texas red, 94
Oak, water, 94
Oak, white, 93
Oak, willow, 94
Orange, trifoliate, 87
Osage-orange, 79

Paper-mulberry, 53
Partridge-berry, 81
Pawpaw, 51
Peach, 89
Pear, Bradford, 91
Pear, common, 91
Pepperbush, coastal, 60
Pepperbush, mountain, 60
Pepper-vine, 49
Pepper-vine, heartleaf, 49
Periwinkle, common, 113
Periwinkle, greater, 113
Persimmon, 64
Pine, eastern white, 86
Pine, loblolly, 86
Pine, pitch, 86
Pine, shortleaf, 86
Pine, Table Mountain, 86
Pine, Virginia, 86
Pinxterbloom, 96
Pipe, Dutchman's, 50
Pipe, woolly Dutchman's, 50
Pirate-bush, 53
Planer-tree, 87
Plum, American, 89
Plum, Chickasaw, 89
Plum, hog, 90
Plum, hortulana, 89
Plum, Mexican, 89
Plum, wild goose, 89
Poison-ivy, 108
Poison-oak, 108
Poison-sumac, 108
Poplar, balsam, 88
Poplar, white, 88
Porcelain-berry, 49
Prairie-redroot, 57
Prickly-ash, common, 115
Prickly-pear, eastern, 83
Princess-tree, 84
Privet, border, 76
Privet, Chinese, 76
Privet, European, 76

Quince, flowering, 58

Raccoon-grape, 49
Rain-tree, golden, 75
Raspberry, black, 101
Raspberry, flowering, 101

Raspberry, gray-leaf red, 100
Redbud, eastern, 58
Red-cedar, eastern, 74
Rhododendron, Piedmont, 96
Rose, Eglantine, 99
Rose, Japanese, 75
Rose, memorial, 99
Rose, multiflora, 99
Rose, pasture, 99
Rose, prairie, 99
Rose, swamp, 99
Rose, Virginia, 99
Rosemary, Cumberland, 60
Rose-of-Sharon, 71
Russian-olive, 65

St.-Andrew's-cross, 72
St.-John's-wort, bushy, 72
St.-John's-wort, cedarglade, 72
St.-John's-wort, early, 72
St.-John's-wort, fivelobe, 72
St.-John's-wort, shrubby, 72
St.-Peter's-wort, 72
Sand-myrtle, 75
Sassafras, 103
Sea-myrtle, 51
Serviceberry, Canadian, 49
Serviceberry, common, 48
Serviceberry, red, 49
Serviceberry, smooth, 49
Silverbell, 70
Smoke-tree, American, 62
Snowbell, American, 106
Snowbell, bigleaf, 106
Snowberry, 106
Snow-wreath, Alabama, 83
Sourwood, 84
Sparkleberry, 111
Spicebush, 76
Spiraea, Appalachian, 105
Spiraea, Japanese, 105
Spiraea, plumleaf, 105
Spire, sweet, 73
Spotted-wintergreen, 58
Spruce, red, 85
Star-vine, bay, 103
Strawberry-bush, 66
Strawberry-bush, running, 66
Sugarberry, 57
Sumac, fragrant, 97
Sumac, smooth, 97
Sumac, staghorn, 97
Sumac, winged, 97
Supple-jack, Alabama, 52
Swamp-privet, 67
Sweetbay, 80
Sweet-breath-of-spring, 78
Sweet-fern, 60
Sweetgum, 77

Sweetleaf, common, 107
Sweetshrub, eastern, 54
Sycamore, 87

Tea, New Jersey, 57
Thorny-olive, 65
Tree-of-heaven, 47
Tulip-poplar, 77
Tuliptree, 77
Tupelo, swamp, 83
Tupelo, water, 83
Twinflower, 77

Upland-privet, 67

Viburnum, Kentucky, 112
Viburnum, leatherleaf, 112
Viburnum, mapleleaf, 112
Viburnum, plumleaf, 112
Viburnum, Rafinesque's, 112
Vine, buckwheat, 53
Virgin's-bower, 59
Virgin's-bower, Catesby's, 59
Virginia-creeper, 84
Virginia-willow, 73

Wahoo, eastern, 66
Walkingstick, devil's, 50
Walnut, black, 74
Walnut, white, 74
Water-elm, 87
Water-locust, 69
White-cedar, northern, 108
Willow, black, 102
Willow, Carolina, 102
Willow, gray, 102
Willow, heartleaf, 102
Willow, prairie, 102
Willow, sandbar, 102
Willow, silky, 102
Willow, weeping, 102
Willow, white, 102
Wineberry, 101
Winterberry, common, 73
Wintergreen, 68
Wisteria, American, 114
Wisteria, Chinese, 114
Wisteria, Japanese, 114
Witch-alder, mountain, 67
Witch-hazel, American, 70
Withe-rod, 112

Yellow-poplar, 77
Yellowroot, 115
Yellowwood, Kentucky, 59
Yew, Canadian, 107
Yucca, 115